The Limits of Matter

synthesis

A series in the history of chemistry, broadly construed, edited by Angela N. H. Creager, Ann Johnson, John E. Lesch, Lawrence M. Principe, Alan Rocke, E. C. Spary, and Audra J. Wolfe, in partnership with the Chemical Heritage Foundation

The Limits of Matter
Chemistry, Mining, and Enlightenment

Hjalmar Fors

The University of Chicago Press :: Chicago and London

Hjalmar Fors is a researcher and teacher in the Department of History of Science and Ideas at Uppsala University, Sweden.

The University of Chicago Press, Chicago 60637
The University of Chicago Press, Ltd., London
© 2015 by The University of Chicago
All rights reserved. Published 2015.
Printed in the United States of America

24 23 22 21 20 19 18 17 16 15   1 2 3 4 5

ISBN-13: 978-0-226-19499-8   (cloth)
ISBN-13: 978-0-226-19504-9   (e-book)
DOI: 10.7208/chicago/9780226195049.001.0001

Library of Congress Cataloging-in-Publication Data

Fors, Hjalmar, author.
    The limits of matter : chemistry, mining, and Enlightenment / Hjalmar Fors.
        pages cm — (Synthesis)
    ISBN 978-0-226-19499-8 (cloth : alk. paper) — ISBN 978-0-226-19504-9 (e-book)
1. Matter—Philosophy. 2. Chemistry—History. 3. Mining engineering—History.
4. Enlightenment. I. Title. II. Series: Synthesis (University of Chicago Press)
    BD646.F67 2015
    117—dc23

2014017926

To Karin, Hedda, and Beata

. . . I heard footsteps behind me
And there was a little old man
—Hello!
In scarlet and grey, shuffling away
—(laughter!)
. . .
—Oh, I ought to report you to the Gnome office
—Gnome Office?
—Yes
—Hahahahaha!
. . .
Ha ha ha, hee hee hee
I'm a laughing Gnome and you can't catch me
Said the laughing Gnome

DAVID BOWIE, "The Laughing Gnome" (1967)

Contents

1

Introduction: The Edges of the Map

Intellectual battles over the nature of reality were fought all over Europe in the seventeenth and eighteenth centuries. They were an integral part of the cultural shift that is usually referred to as the Enlightenment. At the heart of the matter lay a massive change in Western epistemology. The people of this period raised many questions, and found their answers to be different from those that had satisfied their forebears.[1] What are the limits of attainable knowledge? How should we proceed to understand the universe? What kinds of entities and forces may our world contain? The issues were not just philosophical. Debates raged in many arenas. There was the juridical question of whether the law should continue to punish witches who performed malevolent magic. There was the theological question of whether God still worked miracles in the modern world, and whether God and the Devil were aided in their work by legions of angels and demons. And what about the soul? Was it a spark of the divine or simply an innate ability to reason rationally? The debate also raged in physics, natural history, and alchemy/chemistry. How should physics explain phenomena such as action at a distance and telepathic communication of thoughts and emotions? How should natural historians discuss the various beasts—dragons, trolls, gnomes, and the like—that could not be killed, preserved, and displayed in natural history collections?² Should alchemists and chemists abandon their quest for the philosophers' stone, and with it their hopes of transmuting lesser metals into gold?

Central to many of these discussions was the relationship between the realm of spirit (or imagination) and the realm of matter. Perceptions of matter changed radically as many influential Europeans turned their backs on witches, trolls, magic, and miraculous transformations. Whereas matter once had been seen as malleable, transmuting, and ever-changing, it would increasingly be treated as predictable and possible to systematize into stable categories.

This is a book about how the modern notion of materiality was established during the first half of the eighteenth century. It shows how alchemists and chemists contributed to Enlightenment discourse about matter by defining some objects as natural and others as out of the ordinary and probably nonexistent. In doing so, it pins an important epistemological change in European culture to the formation of the modern discipline of chemistry.³ When matter is redefined and given new boundaries, notions of the nonmaterial and of the spirit world change also. Hence, this book takes the debate about the Enlightenment, which has mostly been confined to fields such as the history of philosophy, theology, and physics, into a new arena. As Cyril Stanley Smith remarked in 1967: "until recently, it has been mainly chemists who have contributed to the understanding of the nature of matter."⁴ This book invites not only chemists into the debate, but also assayers, miners, mineralogists, and alchemists, that is, those who knew the most about the rocks, minerals, metals, and mines that comprise the flesh, bones, and internal cavities of the Earth itself.

The Early Modern Point of View: An Introduction

When studying the early modern period, present-day notions of reality should not be taken for granted. A whole range of phenomena that today would be considered supernatural or superstitious were considered epistemically unproblematic by early moderns. That is to say, they had a different view of what reality and God would permit to happen in the normal run of things. Examples of beliefs that were acceptable to most of them, but that are unacceptable to most of us, are that metallic transmutation was viable, that one could use magical techniques for enchantment and treasure finding, and that intercourse with the spirits of the dead and sentient nature-spirits (who were endowed with a form of reason) were not uncommon occurrences. Indeed, our distinction between natural and supernatural phenomena does not at all suffice to distinguish between what early modern people thought that nature would and would not allow.[5] All of the above, as well as many other phenomena, were considered suitable objects of curious investigation by early modern natural philosophers. Not only where they thoroughly discussed, but they were considered important areas of investigation.

Historian of science Lorraine Daston has pointed out that the late seventeenth century delighted in inexplicable phenomena, variety, and surprise. This interest in all things apparently strange can be traced not only in tracts on natural philosophy but also in learned theories on magic and spirits, demonological tracts, court proceedings from witch-craft trials, and ethnographical accounts. There was plenty of room for this kind of inquiry in the realms of physics, chemistry/alchemy, and natural history.[6] Daston claims that toward the turn of the eighteenth century, 1700, interest in these matters was becoming unfashionable. She persuasively argues that the fading of interest in curious phenomena happened through a shift of focus, precipitated by Enlightenment utili-tarianism. As she puts it: "A new ethos of utility replaced the old one of curiosity . . . the stabilization of physical phenomena [was understood] as the necessary, if not sufficient condition for practical applications."[7] Utilitarian outlooks and a search for ingenious inventions, which carried the potential to improve society and yield economic benefits, replaced curiosity.[8]

Indeed, just as Daston indicates, things such as trolls or magic ceased to command the curious attention of mid-eighteenth-century natural philosophers in the way they had a generation before. But there were also strong continuities. When one regards the subtext of enlightened utilitarian discourse, it is apparent that these objects of knowledge con-tinued to have an importance. Increasingly categorized as superstitions outside of science, they were discussed nonetheless.[9] The emerging ide-ology committed to the rationalization and commodification of both nature and culture needed the category of superstition as an object of ridicule. Superstition was opposed to rationality, and the superstitious was one of the major groups (along with artisans/craftsmen and others) that, allegedly, did not show an interest in making knowledge public and useful for the service and improvement of society. In the rhetoric of would-be reformers, useful knowledge, economic reform, rationality, and Enlightenment could be contrasted to allegedly useless enterprises and speculations, stagnation and superstition. This kind of utilitarian rhetoric was equally useful whether one sought to overthrow society, improve it, or simply advance one's own career.[10]

Among the fields of early modern inquiry, the shift toward Enlightened utilitarianism was felt most strongly in the knowledge area
of chemistry. In a very important sense, modern chemistry was born
through this utilitarian transformation of its goals and outlook. During
this period, chemistry transformed itself into a service science, facilitating state control and aiding in industrial development. It gained a place
in society that it has held ever since. Simultaneously the quest for utility,
and for the patronage of manufacturists, led chemists away from their
art's long-term association with transmutation of metals and the quest
for the philosophers' stone. This change was also central to chemistry's
rejection of the strange and the curious.

Mechanical chemistry, as the new breed of chemists sometimes
called their field, allowed the existence of no objects that could not be
isolated and handled in the laboratory. In fact, the connection between
chemistry's rejection of economically useless investigative areas and the
stabilization and description of matter in the chemical laboratory is the
key to understanding chemistry's part in eighteenth-century Enlightenment sentiments. It is the aim of this book to uncover this connection
in the historical material, in order to show chemistry's contribution to
the larger processes that transformed European society during the late
seventeenth and eighteenth century.

The most striking change that chemistry underwent during the first
decades of the eighteenth century was the rejection of gold making, or
chrysopoeia. Before this period, there was no broadly accepted demarca-
tion between alchemy and chemistry—neither theoretically nor when it
came to methods and goals—but in the eighteenth century alchemy and
chemistry split in two. At this point, chemistry was joined to the utilitarian movement as outlined above and remodeled itself as a rational, useful enterprise at the service of society. Alchemy splintered off from the
discourse of utility to which it previously had belonged and was used to
designate the quest for the philosophers' stone, the art of metallic transmutation, and in particular the practice of gold making. As time passed,
it became increasingly identified with irrationality and madness.[11]

To avoid the anachronisms that come with the associations that were
the outcome of this split, in this book I follow the usage of William
Newman and Lawrence Principe in adopting the term *chymistry* to signify the unified pre-1720s tradition of alchemy and chemistry as a joint
enterprise. Where it is necessary to distinguish between chymistry in
general and the specific practice of manufacturing gold or other metals
through transmutation, the term *transmutative chymistry* is used, not
alchemy. Transmutative chymistry is, nevertheless, regarded as an inte-

grated part of the enterprise of early modern chymistry as a whole. From chapter 4, however, which discusses events that took place from the 1730s and onward, the terms *chemistry* and *alchemy* are reintroduced because from about this time these terms begin to be routinely used in a sense roughly equivalent to contemporary usage.

Enlightenment Redefinitions of Nature

To change one's perceptions of nature is to engage in a process with both social and scientific aspects. Indeed, there is no way to separate the two. During large parts of the eighteenth century, the proper stance of the social and learned elite was widely held to be that of skepticism. These groups began to associate continuing belief in several entities and phenomena that previously had been acknowledged as within the bounds of nature, with the lower strata of society, foreigners, and the mentally ill.[12] This attitude comes across clearly in a letter from the Swedish chemist and mineralogist Axel Fredrik Cronstedt, written in 1758 to a friend. In a discussion of alchemy, Cronstedt stated that it should properly be seen as one out of three separate branches of "mystical" (i.e., secret) science, with the others being astrology and magic. As he put it: "[astrology and magic,] the two first parts of [mystical] Physics . . . are out of fashion [and] persecuted, while the last, or alchemy, holds on to its good reputation . . . I am, though, secretly convinced . . . that alchemy is a daughter of magic, who carries a more fashionable dress than . . . her loving mother, and thus can show herself in better company."[13]

After this likening of alchemy to a woman dressing above her station, Cronstedt moved on to ascribe magic and astrology to the lower classes, in particular peasants and the ethnic minorities of Finns and Sami. Alchemy was described as still fashionable among better folk but pursued almost exclusively among the uneducated: belief in these disreputable "sciences" belonged mostly to bygone days or was ascribed to youthful folly. This left the views Cronstedt himself propagated safely in the hands of mature, upper-class men with an up-to-date education.

Thomas Gieryn has proposed that the act of ascribing different beliefs about nature to diverse social groups should be called boundary work and should be seen as a type of contest for credibility. The term designates a rhetorical activity conducted for the purpose of creating a boundary "between science and some less authoritative residual nonscience."[14] In order to create and re-create, "the legitimate power to define the real," it is, Gieryn observes, necessary to engage in a number of social activities that involve a large number of actors, over long timespans.[15] Through these activities, precise scientific arguments are re-placed by broad comparisons between good and bad knowledge. Some practitioners are branded as unreliable, or even fraudulent, while others are considered reliable and honest. Certain ways to represent reality are marked as credible and others as irrelevant, problematic, or wrong. Through these processes, science is placed at the center of a set of posi-tive associations and characteristics. Nonsciences and almost-sciences, on the other hand, are rejected and relegated to the status of unreliable or false pseudo-knowledge. This study uses Gieryn's general framework to explore the interplay between notions of materiality and immaterial-ity among learned practitioners of chemistry and other knowledge hold-ers engaged with the mining business.

Due attention to both publicly held stances and privately circulated texts permits one to pay attention both to the front stage and back stage of public statements of truth. Often, various seemingly noncom-patible theoretical frameworks were used to discuss different types of phenomena, and statements of truth could differ according to contexts. It is necessary to pay proper attention to undercurrents of nonorthodox beliefs in discussions of curious phenomena among the learned.[16] Spirit sightings, miraculous events, magic, the alien, strange, and weird did not just go away with the coming of the eighteenth century. In addition, when such objects were discussed seriously and earnestly, arguments could be surprisingly cautious, and final judgments were often with-held.[17] The study of curiosities, anomalies, and apparent irrationalities never disappeared entirely, and many publicly held views were often paired with secret doubts or qualifications.

A central issue for the study is the social construction of metals as metallic elements, a historical process that has not been widely ac-knowledged in the history of science. Recently, Klein and Spary de-scribed metals as "materials" that were "painstakingly extracted from the earth" and that were uncontested as objects of inquiry.[18] These au-thors also claim that theories of the ultimate structure of matter were of little concern to eighteenth-century "hybrid experts" on materials and had little influence on their practices of experimentation and ob-servation.[19] I disagree. As we will see, pre-eighteenth-century systems of knowledge did not simply assume that metals were raw materials to be collected and transformed into commodities. Metals were not materials in any modern sense of the word prior to the first half of the eighteenth century; they were constructed as such—assigned to that category—through the economizing language of mid-eighteenth-century chemistry. Furthermore, the redefinition of metals played an important role for the creation of chemistry: it happened partially through the merging of chymistry with natural history and the artisanal practice of assaying. Hence, these redefinitions of metals were conducted by precisely such hybrid experts as, Klein and Spary believe, were unconcerned with theories of matter.[20] In this book, the process is inscribed into the story of how chemistry disassociated itself from alchemy in theory, practice, and rhetoric, and in the wider context of mining knowledge, and of Enlightenment thought about the limits of natural knowledge.

Decentralizing the Enlightenment

To understand how chemistry was a part of and contributed to Enlightenment thought, it is necessary to piece together again the shared European natural philosophy and culture of the seventeenth and eighteenth centuries. National interest and the worship of individual genius have depicted intellectual endeavor as separate and particular. In fact the intellectual world of early modern Europe was much more cosmopolitan, integrated, and nonobservant of nation-states than most of us would think. Indeed, the symbiotic relationship between science and nation-states which nowadays often is taken for granted is a product of historic developments that to a large extent played out in the nineteenth century. If, as often has been done, one cuts away parts of the wide intellectual landscape of early modernity, one also understands much less about the remaining parts. This study pays much attention to people, places, and events normally relegated to the sidelines of narratives of Enlightenment knowledge production, for example, artisans, chymical adepts, spirits, witches, and mining administrators. This, I believe, will contribute to the redrawing of the map of seventeenth- and eighteenth-century science. As historian of science Kapil Raj has persuasively argued, not individuals, institutions, or places should be considered centers of scientific change but the cultural contact zones that permit circulation of knowledge in the first place.[21]

This book introduces one of the most productive contact zones of chymistry, mineralogy, and mining knowledge in late seventeenth- and eighteenth-century Europe: the Bureau of Mines (Bergskollegium) of the Swedish state. Sweden's rulers controlled Europe's largest (until ca. 1750) export-oriented iron industry and were highly dependent on the production of minerals. They took a firm grip on this flourishing mining business and actively encouraged mining officials to seek ways to improve it. The Bureau of Mines was turned into a tool for the systematic improvement of mining practices and the pursuit of mining knowledge.

The aspect of the organization of primary interest here is its role in circulation of knowledge throughout early modern Europe, and the changes in natural philosophy, epistemology, and ideology that followed from this circulation.[22] By following actors connected to the Bureau of Mines on their journeys, in their private discussions, and in their published papers, this book reintegrates the natural philosophy and knowledge practices performed at the Bureau of Mines into the European context to which it belongs. In doing so, it highlights an understudied and previously largely ignored actor within the development of late seventeenth- and eighteenth-century learned knowledge as a whole.[23]

Special attention is paid to travelers. Older scholarship often presented knowledge-seeking travelers as going out into the world and coming home with knowledge and objects, as if leaving for an extended journey in order to interact with other scholars and knowledge carriers was no different than going from your house to the supermarket and back again! At a closer look, a much more complex picture emerges. Travelers and traveling are central to the creation and rearticulation of knowledge. Here emerges a narrative of native residents, foreign residents, long-term visitors, and strangers, all engaged in communication with one another. For each type of encounter, a different social dynamics is brought into action that not only changed the distribution of knowledge but also renegotiated relationships and power relationships between the involved actors and the various regions that they eventually made the choice to settle in. Traveling just about everywhere where mining knowledge could be found and oftentimes settling in as longtime students, the officials and auscultators (student-apprentices) of the Bureau of Mines were no ordinary travelers. Although strangers in many of the locales where they arrived, they were backed up by a powerful organization and by domestic patrons. Furthermore they often upheld regular epistolary exchanges with colleagues at home and were expected to produce a detailed travel narrative on their return. As the eighteenth century progressed, they were also increasingly seen, and used, as resources by those whom they encountered en route.

The officials of the Bureau were busy travelers who gathered knowledge in particular in the Holy Roman Empire, England, and the Netherlands. The book focuses on travelers, transfers of knowledge, and on long-distance exchanges of intellectual influences, as narrated in travelers' reports, correspondence between practitioners of natural philosophy, published books, and circulated and private manuscripts. Official documents such as protocols and laws and regulations are only discussed as complementary material, and only when they cast light on the subject at hand. It is precisely the focus on the Bureau's role in transnational knowledge exchanges that opens up new perspectives on the relationship between science and the central philosophical debates of the late seventeenth and eighteenth centuries.

The study links together seventeenth-century Swedish aristocrats and German artisans. It discusses places such as London and Leiden as continuing sources of inspiration for travelers, but it also shows how travelers transformed and used what they learned there, after they had returned home. Linking pieces such as these reveals patterns of knowledge circulation and cultural exchange and shows a deep affinity between regions on a cultural level. I hope the book will deepen our understanding of Enlightenment thought as something bigger and much more interesting than, to use the words of Simon Werrett, "Anglo-French reason, progressive politics and liberty."[24]

Werrett argues that alleged peripheries, such as Russia, tend to be treated as partaking in European science only insofar as scientific works produced there "cross the borders to the West, to Germany, France and England, and conversely, only attain Enlightenment when the best German, English and French thought reaches Russia."[25] This book presents a similar argument concerning the knowledge production at the Bureau of Mines. It proposes that centrist historiographical accounts are constructed by de-localizing and disembodying the sciences of the periphery. Through these reconstructions, the alleged peripheries are denied all real agency in, or influence over, the alleged centers. Centrist historians imagine that foreign knowledges were of little consequence before they were translated into a central language and integrated into a central knowledge community. There is, however, a problem with historical narratives that assume travelers did little of value once they returned home from, say, Paris or London. Or narratives that assume books published in, say, Uppsala or Saint Petersburg had no or little influence before they were translated into English or French. In the end, the centrist approach feeds credibility into outdated narratives centered around the concept of national sciences.[26] These narratives have done a lot of harm in that they conceal communication between geographically dispersed knowledge communities, which, arguably, is a significant driving force behind scientific and technological change.

All of these considerations have bearings on this study in a concrete sense. This book argues that the chemists at the eighteenth-century Bureau of Mines made a significant contribution to the modern European view of reality. They presented novel interpretations in matter theory as well as the modern notion of the chemical element. Historians of physics have long been concerned with how physics has been a force of reality change and reality reconstruction. Strangely enough, historians of chemistry rarely tackle this important issue head on, arguably because they have created a largely centrist historiography. That is to say, the Bureau of Mines and other similar milieus represent a significant lacuna in our knowledge of eighteenth-century chemistry and mineralogy. To learn about what happened when these disciplines emerged during the seventeenth and eighteenth centuries, one has to study a wide range of places, as well as original sources about artisans and *curiosi* such as assayers, apothecaries, and chymists. One cannot be content with discussing over and over again the relative importance of the familiar group of well-known heroes that is routinely called up in older histories of science.[27]

The contribution of the Bureau's chemists to systematic classification and to chemical analysis of minerals has been noted by a number of international scholars beginning already in the eighteenth century and continuing to this day. Georg Brandt's 1735 discovery of cobalt and Axel Fredrik Cronstedt's 1751 discovery of nickel are rather well known, and so too is Cronstedt's mineralogy of 1758.[28] Nevertheless, despite the many available works in a number of languages, no historians of science, or of Enlightenment, have troubled themselves with the wider contexts into which these so-called contributions were inscribed.[29]

Brandt and Cronstedt and other Bureau chemists sought to redefine nonbiological matter—the mineral realm—as consistent of solid and tangible objects that could be predictably manipulated. Between 1730 and 1760, several minerals were defined as previously unknown species of metal at the Bureau of Mines, that is, they were discovered, in the sense that they were isolated, described, and presented as novelties to the international community of chemists. Here we find, fully articulated for the first time, the modern concept of the chemical element.

This then, is the book's main contribution to the history of science: it situates the concept of the chemical element in its original context of Enlightenment boundary-work around nature. It does this by outlining how mineralogical chemists brought together knowledge from a wide variety of sources, with the aims to, first, reorganize mining knowledge along utilitarian lines; second, enlist natural philosophy at the service of the state's fiscal interests and third, erect social boundaries toward forms of knowledge deemed unacceptable or useless. Finally, the book also shows how an array of entities and phenomena previously conceived of as natural were defined as outside of nature through this process.

But what does it really mean to say that a group, place, or organization is more important than another to the development of an intellectual endeavor? Eighteenth-century enlightened discourse, which tended toward universalism, was uneasy with such distinctions. And so should we be. In view of the untenability of the centrist mode of research, this study does not want to establish the centrality of the Bureau but seeks to highlight how the Bureau's apparent centrality was a consequence of its remarkable ability to function as a contact zone between different groups in society, as well as between different geographical regions.

Chapter 2 begins with the Stockholm witch trials, in which the doubts cast on the testimonies of the accusers cast further doubt on the power of witchcraft. Through these trials, we also gain insight into the intellectual world of Urban Hiärne, the first major chymist to be active at the Bureau of Mines. Chymistry, or rather alchemy, has often has been singled out as an occult and suspect enterprise. The chapter makes clear that interest in curious natural—as well as otherworldly— phenomena was an integral part of several other early modern forms of knowledge, such as physics, natural history, and law. By establishing the facts and beliefs held about nature in late seventeenth-century Europe, the chapter sets the stage for an informed discussion of elite skepticism in the eighteenth century.

Chapter 3 begins with an introduction of late seventeenth-century chymistry and shows how actors in the Holy Roman Empire and the European north perceived it as an economically oriented and useful activity. It introduces the environment of the Bureau of Mines and presents the story of how transmutative chymistry was established and came to flourish there. Central to these developments was active circulation of knowledge about crafts and chymistry between Sweden and the mining areas of the Holy Roman Empire, in particular Saxony. It came about through an active policy on part of the Bureau, and through the joint effort of Hiärne and his closest disciple Erich Odhelius. Here I also describe how transmutative chymistry, through accumulated evidence, came to be dismissed as lacking utility.

Chapter 4 investigates the influence of mechanical philosophy on chymistry during the first half of the eighteenth century. The first section of the chapter discusses the journey to England by Odhelius and maps out how he became an Anglophile, and an advocate of the new experimental philosophy. I then shift attention to the university town of Uppsala and investigate how the Bureau, through the mechanic Christopher Polhem and others, became a center for Uppsala-inspired late Cartesianism, mathematics, and engineering. I then again turn to London and investigate how the loosely connected group around the Bureau and at Uppsala were influenced by, and reframed, recent English developments. In particular,

I investigate the work and journeys of Polhem's protégé and assistant, the young Emanuel Swedenborg. Moving on from London and Uppsala, I turn to Leiden in the Netherlands to trace how Herman Boerhaave's reconceptualization of the new experimental philosophies and methods gained a decisive influence over the chemistry of the Bureau of Mines.

Chapter 5 investigates the interconnectedness of mining knowledge in the wider German and Swedish cultural sphere, and how the officials of the Bureau began to distinguish themselves from their counterparts in the Holy Roman Empire. It outlines how the chymists of the Bureau articulated a new type of mechanical, mineralogical chemistry by integrating chymistry with assaying practices, and with natural history. This new thinking thoroughly transformed the production of knowledge at the Bureau and established the foundations of a new science of mining, devoid of what now had come to be regarded as superfluous speculation, and with a new theoretical and methodological foundation in chemistry and physics/mechanics.

Chapter 6 studies how the new brand of chemists dealt with those objects of knowledge that were now out of bounds. Through active boundary-work, belief in, for example, trolls and the efficacy of magic and transmutation was transposed onto foreigners and ethnic minorities, women and artisanal craftsmen. The chapter also examines some beliefs and practices associated with spirits and subtle matters that remained within the permissible discourse of eighteenth-century natural philosophy. We will also witness a great battle in which those natural philosophers who gradually had built up their skepticism over decades finally emerged. Regardless of whether the doubts concerned magic, alchemy, religious enthusiasm, the supernatural, or simply metaphysics, they were now put forward openly. The eighteenth century's turn toward skepticism, however, was a highly localized phenomena, and as such limited to certain individuals, groups, and social exchanges. The study ends at a cultural turning point, at which self-proclaimed enlighteners found the courage to voice public skepticism. By the 1760s enlightened debates in Protestant Europe began to enter a more assertive and aggressive stage. These developments, in conjunction with the spread of Kantian philosophy, would render obsolete many of the issues and considerations that engaged the protagonists of this book.[30]

A Note on Epistemology, Terminology, Translations, and Conventions

In a work that deals with matter, spirits, magic, trolls, and the social construction of knowledge, it may be necessary to delineate the book's epistemological foundation and the author's take on the issue of the truth of scientific facts. To put it bluntly: do not expect either a story of the triumphant rise of Enlightenment reason or a Weberian narrative of the gradual disappearance of the enchanted world and its replacement with rationality and order. I am not concerned with whether one type of knowledge is superior to another. The book's underlying, albeit central, aim is to study *under which social conditions* certain forms of knowledge are perceived as better than others. Furthermore, a symmetrical style of explanation is used, by which I mean that similar explanations are used to explain the successes or failures of all relevant groups of actors.[31]

Just as this is not a book on the triumph of Enlightenment reason, it is also not a book on the rise of radical skepticism. The process examined in this book should not be described as a mounting scientific skepticism of various superstitions (which was, as it were, undertaken by single, heroic individuals ahead of their time). This book discusses a series of changes of intellectual positions, one after the other, that constituted valid answers to particular problems and that arose out of particular contexts. Taken together and in hindsight, these changes chipped away at the old block of inherited epistemology. They were, therefore, in a sense, part of a historical process of change, which would culminate in what has been called the Scandinavian and northern German radical Enlightenment of the 1760–80s. But earlier positions should not be considered mere forerunners of later ones. Many positions that were presented to the public in the 1760s would have been considered by most as no more than affronting rudeness in the 1710s. Both earlier and later positions were developed in contexts that permitted them, and that were at least in part ready to welcome them. If one seeks to commit "the ultimate incivility" as Steven Shapin calls it, namely, "the public withdrawal of trust in another's access to the world and in another's moral commitment to speaking the truth about it," then the ground needs to be well prepared. Otherwise the skeptic simply will be expulsed from the community that holds the beliefs that he so blatantly distrusts and disrespects.[32] In a sense, this book is about breakdown and reconfiguration of cognitive and moral order: a view of Enlightenment from the other side of the mirror.

There are several terms used in this book that have undergone a change of meaning through the centuries. Generally, whenever possible, I use terms in a sense similar to that of the actors. Instead of *science* and *technology* I use the more generic term *knowledge*, which would have included learned bookish and systematic knowledge as well as the skills of craftsmen and artisans. The term *science* is used sparingly. Either it is used in an easily recognizable modern sense, or as a translation of its early modern close equivalent, as used by actors, and denoting a systematic knowledge that also had practical applications and that was practiced by high-status individuals. The term *magic* too is used as an early modern actors' category. Just as I discuss the splitting up of chymistry into alchemy and chemistry, I follow the transformation of the meaning of magic through the studied period. In the beginning of the period, interested scholars and learned magicians usually held magic to be within the bounds of nature, and as acting through and within laws of nature. Magical knowledges and practices were primarily contested on moral and religious grounds due to their association with, for example, heresy, paganism, or the Devil. (For a contemporary equivalent, think *Nazi science* rather than *occultism* or *new age*.) Toward the end of the period, the term was beginning to take on its modern connotations of unscientific or fake knowledge. Furthermore, the book also distinguishes between witches, or Sabbath witches (who rarely could be found outside of the courtroom context of forced confessions, and in folktales and fairy tales), and the second-sighted, as well as practitioners of various types of magic used for healing and harm and for the benefit of themselves and their local society, all of whom are designated as cunning. The term *trolls* designates beings not dissimilar to the tailed trolls of contemporary popular culture. In early modern Scandinavia, however, trolls could also be described as beautiful, that is, sexually alluring, and well clad. Furthermore, the category of troll was often conflated with various keeper entities (or keeper spirits) associated with certain places: the keepers of the mountains, mines, forests, lakes, and streams. Here no clear distinction is made between these two categories of beings. I do not speak much about the many other entities known to Scandinavian folklore, as they rarely were associated with mining and precious metals in the way that trolls were.[33] Regional differences, such as the distinction between the Scandinavian troll and the German kobold, are discussed when relevant.

The Swedish and German term *bergsman/Bergmann* (literally, mountain man) has no English equivalent and is therefore used with no translation. It signified individuals who possessed a practical knowledge of mining, including skilled mineworkers, smelters, shareholders in mines, and owners of small-scale smelting works, as well as mining officials.[34] The Swedish mining authority Bergskollegium is translated as the Bureau of Mines, while earlier research often has used the translation Board of Mines. (Seventeenth- and eighteenth-century texts often translated the name of the organization as the College of Mines.) Older translations have been abandoned for the novel term the Bureau of Mines. The term Board emphasizes the organization's collegial structure at the top level. As in the *Bergämter* of the mining states of the Holy Roman Empire, all major decisions at the Bureau were discussed and voted upon by a governing *board*. It consisted of a small group of senior officials bearing the titles Assessor and Councilor, and a politically appointed president. The vast majority of Bureau employees did not, however, have a seat and speaking rights (Ger., *Sitz und Stimme*) on the governing board. Hence the term Bureau is used to designate the whole of the organization and to convey an idea of its size and role as a central mining authority staffed by a number of officials with specialist functions, and endowed with a number of local branches. I should also mention that Germany is sometimes used as shorthand to designate the predominantly German-speaking lands of the Holy Roman Empire. This corresponds to common usage in several European languages of the time, as well as to usage among a fair number of seventeenth- and eighteenth-century speakers of German.

All translations into English are my own, unless otherwise stated. I have tried to render translations as true to original wording and sentence structure as possible, although this practice may at times have yielded slightly unwieldy sentences. Some originals are written in a mixture of several languages (i.e., Swedish, German, and Latin, sometimes in the same sentence); these have been translated into English without any markers concerning the language of the original.

2

Of Witches, Trolls, and Inquisitive Men

Prologue: The Stockholm Witch Trials

Two accused witches were burnt in Stockholm in the summer of 1676. Their names were Malin Mattsdotter and
Anna Simonsdotter Hack. The latter was decapitated and
burnt, while Malin Mattsdotter was burned alive. Both
women were sentenced to their deaths by a special commission, authorized by King Charles XI to investigate a
severe outbreak of witchcraft accusations in the city.[1] The
commission on witchcraft represented the cream of Stockholm's learned and official world: clergymen, lawyers, and
men of the state. Among them, representing the medical
profession, was Urban Hiärne, physician, chymist, and
fellow of the Royal Society of London.[2] The commission
would investigate whether innocents, in particular children, had been put into a state of trancelike sleep. Then it
was said that they had been stolen away for the witches'
Sabbath at a place called Blåkulla. Sometimes they had
arrived there riding on the back of a neighbor, sometimes
on a stick. There was for example the servant girl Annika
Hinrichsdotter, twenty-one-years-old, who claimed she
had ridden to Blåkulla on several people, among them
her master, Jonas Myra. At Blåkulla the witches and those
they had taken with them met with the Devil. One of the
accused women was said to have slept with him.

Initially, the commission found no reason to doubt the events as related to them by the witnesses. All participants agreed that illegal and morally illicit events had taken place. Execution of the accused proceeded in an orderly fashion. But as the case expanded and more women were accused, doubts began to surface. One question concerned the epistemological status of the experiences narrated to the commission. Where was this Blåkulla? Was it an illusion fashioned by the Devil or a delusion in the minds of crazy people? Was it a physical, material space or a semispiritual realm? Most alternatives were to be considered by the commission. In any case, it was not a specific physical place but located at different places, for example, in a garden in the city. It was also not clear whether these events happened physically or in the spirit.

The commissioner Magnus Pontinus, a clergyman, was clear in his opinion. Many of those who were taken simply fell into a trance and were fooled into believing themselves transported "through Satan's illusions." Pontinus did not deny that there were known cases in which victims had been taken away in their body and replaced by some illusory shape fashioned by the Devil. This case was, however, not one of them. In the accounts of the victims and accused, opinions were divided. One indication that someone had been taken was that their sleeping bodies could not be roused; hence, it seems that bodies were believed to stay behind in a trancelike sleep. According to the testimony of an accused witch, it was true that she had been to Blåkulla, but she had been there in a passive state. Another said that she had been brought there, taken out through her keyhole. On the other hand, those who had been ridden testified of bodily pains and aching limbs the day after their use as steeds, indicating that they had traveled to Blåkulla in the body.[3]

Doubts began to surface only after the hearing of one Karin Erichsdotter. Her niece, Brijta Jonsdotter, was to have testified before the court but instead had escaped the city. Before running away, she had confided to her aunt that her abduction to Blåkulla had been no more than a dream.[4] The testimony of Brijta Jonsdotter's aunt predicated a remarkable turnaround in the opinions of many of the commissioners. Eric Noraeus, a clergyman, stated that he refused to condemn any more people to death unless he was presented with significantly better evidence and information.[5] In September, Hiärne, the physician, summarized his views on the trials in a written statement. He emphatically did not believe in the stories told before the commission. The witnesses had simply populated the world of Blåkulla with grownups and children from their own environment. The parish clerk rang the bells, the baker made bread, a tailor's son made clothes, a carpenter's son made cabi-

FIGURE 1. German print from 1670 showing how children were spirited away to the Sabbath by witches from the village of Mora in Sweden, as well as the burning of condemned witches. Reproduction by the National Library of Sweden, call number KoB Sv. HP C.XI A.8.

nets, and so on. Hence the stories were to be dismissed as inconsistent and ridiculous.[6]

Following the lead of Noraeus, the commission deemed that a third accused witch, Margeta Staffansdotter, was innocent on all accounts. The commission now began to ponder what to make of the accusers. Had they not been seduced by Satan, the Prince of Lies, to lie before the commission so that it would condemn innocent women to death? The commission explained to the servant girl Agnis Eskillsdotter that she should take back the "lies she had hatched in the service of Satan," and Kerstin Michelsdotter, another girl who also kept to her story, was reproached because she "allowed Satan to blind her."[7] Lisbeta Carlsdotter, who was deemed to have been one of the driving forces behind the accusations, was punished. She was accused of being a murderer, a perjurer, and a "corrupter of the souls of innocent children." The commission stopped just short of accusing her as a witch. Herr Jwar, one of the commissioners, believed that she had been possessed, both in spirit and body, and that she should be beheaded. Noraeus too thought that she had been assailed and perhaps conquered by the Devil, but that it was difficult to say anything certain about the matter. He suggested that she be whipped and driven out of the city. The commission went with Jwar. Lisbeta Carlsdotter, two other women, and a teenage boy were condemned and beheaded as liars, murderers, perjurers, and corrupters of the souls of the innocent.[8] The witch craze had now died away in the city, and the commission discussed how they should conclude their business. It was suggested that the prayer about the rage of Satan that was read from the pulpits of the city's churches should be changed. It was important that the populace should not remain under the dangerous delusion of children being taken away to Blåkulla. The pious Noraeus reminded the commission that a thanksgiving should be said in all churches "for the Merciful enlightenment, that God has given in this matter."[9]

Throughout the proceedings, Hiärne had been partly absent and mostly silent. Perhaps it was his autumn wedding plans that kept him busy elsewhere. But were there other, deeper reasons for his silence, having to do with his own interests and involvement with morally spurious systems of knowledge?

: : :

Witchcraft accusations and the weight they were given as juridical evidence serve as a powerful reminder that people of the late seventeenth century defined reality quite differently from most who live in the early twenty-first century. The early modern body was subject to many influences. Not only could it fall ill and become the victim of sudden physical violence. It could also be bewitched and shape-shifted, even squeezed into a keyhole. Epistemological claims about human bodies also contained claims about the material world. It too was fluid and changing. Material objects too could be compressed, expanded, made to disappear and reappear, and affect other objects over vast distances. This is not to say that matter did not behave predictably in many contexts. Chymists, assayers, and other artisans had a stable repertoire of techniques at their disposal to manipulate matter in order to extract essences or valuable

metals, or to create works of art. The bottom line of artisanal and chymical work was metamorphosis, however: objects could be transformed
from one kind into another. Lead or mercury could become gold. Chymistry as conducted in the late seventeenth century and early eighteenth
was an art of transformation. Many considered it dependent on the
grace of God. Hence it was integrated into a wider worldview, carried not so much by natural philosophy as by theology. This chapter
provides a broad outline of some late seventeenth-century learned and
popular discourses in order to establish a basis for the discussions in the
chapters to come. It establishes contact zones between learned theories
of matter, learned and popular theories of magic, chymistry, and the
crafts and lore of miners and smelters. More specifically, Urban Hiärne
serves as our guide through three realms of knowledge: that of magic
and the relation between matter and spirit; that of curious natural philosophy; and that of the influence and importance of trolls and keeper
entities over the business of mining. Through Hiärne, who was head of
the Bureau of Mines Laboratorium Chymicum between 1683 and 1720,
this chapter also sketches the intellectual background to the institutionalization of chymistry in the Bureau of Mines environment.

A Curiosus in the Commission: Urban Hiärne

An artistic and learned man in his midthirties, Urban Hiärne was both
a chymist and natural philosopher, and a poet, playwright, and artist.
In the final decades of his life he would stand out as an aged doyen of
Swedish medicine. He would become the most influential medical doctor in the Realm, as well as the country's most famous chymist. His titles
would include president of the Royal College of Medicine (Collegium
Medicum), personal physician of the queen dowager, and leader and
director of the Royal Chymical Laboratory in Stockholm, a part of the
state's Bureau of Mines. He also served as an assessor, and at times as
vice president of the Bureau.[10] "The foreigners say that there are no
other learned men in Sweden, except Rudbeck and you, Herr Doctor"
claimed the young Erich Odhelius in a 1686 letter to Hiärne.[11] Although
he flattered Hiärne, who was his patron, Odhelius was not stretching
the truth. Hiärne and Olof Rudbeck the elder, professor of medicine in
Uppsala, were the most renowned Swedish men of learning in the late
seventeenth century, and it is doubtful if anyone except the two had a
reputation outside of the country's borders.

   Why did this man of outstanding learning dismiss the stories told
before the commission on witchcraft? The epistemological issue at hand

for Hiärne, as for the other commissioners, concerned the influence and power of the Devil. Could the Devil work change in the material world, or was he limited to act upon the human mind itself? That is, did the Devil operate directly in physical reality, or was he limited by God to work through deception, illusion, and other mind games?[12] According to Hiärne, the Devil wanted to appear immensely powerful because he wanted us to lose our faith in God. But the Devil's power was tightly bounded by the Creator. The events in Stockholm, therefore, were due to other reasons. Hiärne admitted that Satan sometimes involved himself in the affairs of men, but he did not believe there was evidence that Satan really interacted with witches.[13] The fact that Hiärne denied power to the Devil, however, did not necessarily mean that his views of the bounds of the natural and material world coincided with ours.

Seventeenth-century learned acknowledged the existence of several types of magic.[14] Theorists of magic usually distinguished between divine magic, natural magic, and diabolical magic. Divine magic, or works considered miracles, was the preserve of God, his angels, and his chosen prophets and saints. Because the all-mighty God was its direct source, divine magic could completely suspend all ordinary laws of nature. Natural magic, on the other hand, was a phrase that covered all apparently marvelous effects and phenomena that could happen or be made to happen naturally, but that at present could not be explained through ordinary reason. Diabolical magic was performed by devils and demons on behalf of evil men and women who had made compact with them in order to get access to their powers. Many theorists assumed that diabolical magic was a form of natural magic that seemed different and vastly superior to the latter. The main difference was that diabolical magic was performed by ancient and cunning spirits who had spent thousands of years perfecting their magical arts. In this highly moralized view of natural phenomena, divine magic was considered an act of grace, natural magic was encouraged or at least permissible as long as it was not performed for evil purposes, whereas demonic magic, of course, was forbidden.[15]

There were many renowned apologetics of natural magic in the sixteenth and seventeenth centuries, and a number of interpretations and distinctions. In a work published in 1625, the French historian of magic Gabriel Naudé, librarian of Cardinal Mazarin, distinguished among four types of magic. Apart from the three just discussed, Naudé also acknowledged the existence of theurgic or white magic, that is, wonder working under the guise of false, that is, non-Catholic religions.[16] Naudé explicitly stated that natural magic was to be distinguished from the others. While the other forms of magic concerned themselves with celestial

or infernal spirits, natural magic was nothing more than practical physics, "as Physick is contemplative Magick." Indeed,

> whatever the most subtle and ingenious among men can per-
> form, by the imitation or assistance of Nature, is ordinarily com-
> prehended under the name of Magick, untill such time as it be
> discovered by what wayes and means they effect those extraor-
> dinary operations. Of this we have an example in the invention
> of *Guns* and *Printing*, and the discovery of the new world; the
> people wherof, thought at first sight, that our ships were made
> by Magick, our vaults & arches by enchantment, and that the
> Spanyards were the Devils that should destroy them, with the
> thunder and lightening of their Arquebuzzes and Guns.[17]

Chymistry was a part of physics and of natural magic. Well-known
chymical miracles such as fulminant gold and the "Vegetall Tree" were
products of "man's wit."[18] According to Naudé, the famous authors of
chymical treatises, such as Geber, Raymundus Lullius, and Arnoldus de
Villanova, were really learned men and physicians (i.e., medical doctors).[19] Even the infamous Swiss physician Paracelsus did not deserve to
have his name in the catalogue of magicians. As Naudé put it, "though
[Paracelsus] . . . might justly be condemn'd as an Arch-heretick for the
depravednesse of his opinion in point of Religion, yet do I not think he
should be charg'd with Magick."[20]

The main aim of Naudé's distinctions was to separate permissible
forms of magic from forbidden ones. Put another way, he and many of
his contemporary theorists sought to create a space for curious inquiries
that was free from the taint of witchcraft. A certain class of magical
phenomena—natural magic—was redefined as physics and chymistry.
This redefinition separated natural magic from discourses primarily
concerned with religious morality and opened a new field for curious
inquiry. Indeed, one could even say there was a certain late seventeenth-
century rage for curious phenomena associated with magic, spirits, and
the knowledge of cunning folk. In England it was represented by men
such as Joseph Glanvill and Robert Boyle. On the one hand, they regarded these phenomena as terra incognita to be studied with the help
of the tools of the new natural philosophy. On the other, they felt a need
to vindicate phenomena associated with spirits and religion by searching
out empirical evidence in their support.[21]

The reason for the latter was that not all early moderns agreed with
magic's apologists. In particular the Cartesian natural philosophers

were of another opinion. Descartes's philosophy proceeded from the notion of a radical dualism between spirit and matter. Cartesian mechanism contained within it a serious challenge for the notion of a spirit world that was accessible to human reason and inquiry. Although Descartes did not deny the existence of a spiritual world over which God presided, there was preciously little interaction between the worlds of matter and spirit. God interacted with his Creation only through a single gate, and this gate was the pineal gland of the human body. Thus there was no need for any intermediary realms. All movement and change in the physical universe were purely material: everything consisted of corpuscles of matter interacting with other corpuscles in a constant flow of motion and friction. Using the terminology of matter and motion, Cartesian materialism could explain the formation of stars and planets and the reason for their movements in the sky. It could also explain the movement of bodies on the earth, as well as the reasons why they underwent changes.[22]

In the second half of the seventeenth century, Descartes was widely perceived as a great synthetic genius. It was he who had summarized and transformed into a philosophical worldview the mathematics, astronomy, and physics put forward by men such as Nicolaus Copernicus, Johannes Kepler, and Galileo Galilei. In later accounts, pervasive to the present day, Isaac Newton is often given a role similar to the one that the seventeenth century assigned to Descartes. Descartes had spent the last few months of his life at the court of Queen Christina and had died at the royal castle in Stockholm in 1650.[23] Yet it was not until fourteen years later that his teachings began to cause a stir at Uppsala, Sweden's major university—which was perhaps strange, as Uppsala was located just north of the capital, less than a day's journey by boat or on horseback.

In the autumn of 1664 a professor of Greek, Martin Brunnerus, wrote a dissertation attacking Descartes and the mechanical description of reality that lay at the heart of Cartesianism. The small group of Uppsala Cartesians responded quickly, led by the professor of medicine Petrus Hoffwenius, seconded by his colleague, the renowned anatomist and natural philosopher Olof Rudbeck the elder. Hiärne, at the time a twenty-three-year-old student who had arrived to study in Uppsala in 1661, was also part of the Cartesian camp. It was announced that Hiärne would perform the public defense of a dissertation authored by Hoffwenius.[24]

The title of the work was innocent enough, *Artis Medicinalis Parvae Exercitationes* (Short treatises on the medical art). Attached to the dissertation, however, was a corollary titled *Mantissa Physica*, in all likeli-

hood written by Hiärne himself. This text was a compendia of the basic
principles of Cartesian physics and was the first time that an overview
of Cartesian matter theory was publicly presented at Uppsala University. The theologians were not amused. Perceiving Cartesianism to be
a threat to the authority of Scripture, they made sure that the public
defense was postponed indefinitely. The controversy, however, would
continue with more or less intensity until 1689, when King Charles XI
intervened and outlawed philosophical speculation that contradicted
the Bible.[25] It would be another eighty years before a critical, public,
philosophical and religious discussion would again gain momentum in
the Swedish Realm. Controversial philosophies would be spread and
discussed, of course, but they would be proposed cautiously. Religious
censorship and the threat of legal action from religious authorities posed
a real risk for natural philosophers, as well as for citizens in general,
throughout the eighteenth century. It curtailed open discussion of all
matters associated with religion and philosophy.

Hiärne moved on. He left the university in 1666 and must at that
point have been a radical Cartesian, although later events in his life
made him revise his position thoroughly.[26] His time in Uppsala had
gained him valuable patrons in the Swedish elite. Some years later, we
find him traveling extensively in Europe with the financial support of
Count Claes Tott, general governor of Riga. He visited London in 1669
and was made a fellow of the Royal Society. He then moved on to spend
almost three years in Paris. Soon his patron Tott became Sweden's am-
bassador to France and joined him there. Hiärne was in a position to
mix freely with the learned of the city. He visited anatomical and surgi-
cal demonstrations, and was permitted to be present when his patron
received an audience with the king at Versailles.[27] Hiärne also studied
chymistry. The city was one of the greatest centers of this art in Europe.
His studies began in the shop of Christopher Glaser, and from there he
moved on to work under the supervision of two other chymist. The first
was called la Ferte, and the second was an Italian called Praciani. Late
seventeenth-century chymistry was both laborious and costly, requiring
specially made furnaces and other equipment such as glassware, as well
as expensive chemical materials and large amounts of charcoal. Hiärne's
eagerness to learn is confirmed by his assurance that he spent most of
his money on his chymistry and on payment for his chymical teachers
while in Paris.[28]

In 1674 Tott died unexpectedly. As he had no children, his responsibility as patron to Hiärne was inherited by older relatives. These were
the treasurer of the Realm (*riksskattmästaren*) Sten Bielke and the presi-

dent of the Supreme Court (*hovrättspresidenten*) Bengt Horn. Hiärne now had two formidable patrons at the center of political power.[29] He then moved to Stockholm, where he became a physician to the rich and influential. Thinking that his future was now secured, he married the young Maria Svan on October 7, 1676.[30] This was the same autumn that he participated as a commissioner in the Stockholm witch trials.

It is not known how or when Hiärne turned away from Cartesian philosophy.[31] Perhaps it was meeting with the Paris chymists and physicians, or with other knowledgeable men whom he encountered during his European travels, that made him change his mind. But is clear that the materialist implications of Cartesian philosophy never really had suited him. When he returned from his European travels, Hiärne seems to have already embraced a different worldview, one far more compatible with belief in witches than the Cartesianism of his youth. He expressed it through Paracelsian and Neoplatonic philosophy. These would be his steady companions for the rest of his life. In 1709, he named Paracelsus as one of the three foremost philosophers, together with Pythagoras and Plato. He then added that the three of them agreed on most issues. [32]

The Neoplatonic worldview acknowledged the existence of multiple orders of beings and forces, made out of various subtle or gross forms of matter. The most subtle matter was the clear and shining light emanating from the Divinity, but as distance from God increased, matter became darker and coarser. The bodies of the material world were made out of the coarsest stuff of all. Paracelsus, or to use his full name, Theophrastus Bombastus von Hohenheim (ca. 1494–1541), was the author of one of the most important formulations of Neoplatonic philosophy of early modern Europe. Paracelsian natural philosophy became one of the dominant schools of thought in early modern Europe, exerting great influence over medicine, chymistry, philosophy, astrology, and theology, as well as a number of other areas.[33]

Hiärne's philosophy was not just lofty thoughts gathered from books but was also grounded in the folk beliefs of his childhood. He was born in 1641, in Ingria (Ingermanland), a sparsely populated eastern border province of the Swedish realm. His hometown, Nyen, specialized in exporting tar from surrounding forests and was perilously situated at the outlying and slightly neglected fortress of Nyenskans.[34] Today both are gone. The site where they stood is now occupied by a suburb of the Russian metropolis Saint Petersburg, founded sixty-three years after Hiärne's birth.[35] The son of a protestant vicar, Hiärne came from a family characterized by an openness toward second sight and prognostication. Hiärne later noted in one of his manuscripts that there was a little man-spirit in his childhood home, who only his mother could see and who sometimes appeared to give her advice or warning. And when his father had felt that death was upon him, he had gathered the family together and predicted that great change was coming to the Realm of Sweden and that the family's hometown of Nyen would be destroyed. As Hiärne noted many years later, this prediction occurred: Nyen was burnt, and most of its inhabitants were slaughtered when the Russians invaded in June 1656.[36] Hiärne and his family escaped, but with the head of the family dead and their home destroyed, they were now refugees. Two years later Hiärne, now aged fifteen, arrived in Stockholm alone and almost penniless. But he had resources of other kinds. He had learnt the basics of logic, rhetoric, poetry, and grammar, as well as Latin and Greek, at the trivial school of Nyen.[37] Soon he had secured a position in Stockholm as private tutor and made the decision to study medicine in Uppsala.[38]

Prognostication held an important place in Hiärne's personal experience and life story. In his autobiography Hiärne tells of how he first met his future wife in a dream he had had in the German town of Cologne. Initially Hiärne had nurtured plans to secure his financial future by marrying into a rich family, but the dream told him he would marry a young and virtuous but poor woman. Some years later, in March 1676, he spied a girl on a street in Stockholm whom he immediately saw was meant for him. Hiärne followed her, noticed which door she entered, and after making inquiries, learnt that her name was Maria Svan. She was a person of standing, a niece of the famous architect Nicodemus Tessin the elder's wife. But she was completely without means. Hiärne nevertheless asked for her hand in marriage and received her consent. In this way, he stated, everything happened exactly as predicted by the dream in Cologne.[39] An extraordinary event also foretold the parting of the two. When an owl hooted three times outside of Hiärne's window in November 1690, Hiärne took it to be an omen. A month later, his wife died of pneumonia.[40] It is clear that Hiärne framed important and significant events in his own life in relation to invisible and benign forces. The "little man" spirit who could be seen only by his mother had appeared to give instructions when his brother was seriously ill. His father had foretold the destruction of his hometown, and benign forces had intervened to make sure that he married the right woman.

We have seen how Hiärne's worldview and knowledge were compatible with beliefs in magic and invisible beings, which were common in

early modern culture. But that these beliefs were common did not mean
that they could be openly expressed. Hiärne's actions on the commission must be considered in the light of the dangerous association of such
beliefs with witchcraft. The first systematic witchcraft persecutions in
Sweden began toward the end of the sixteenth century. They coincided
with Charles IX's continuation of the Reformation, enacted through a
heavy-handed enforcement of Lutheran orthodoxy. Influenced by foreign examples and in particular German court practice, courts dealing
with cases involving accusations of witchcraft increasingly made use of
torture and the death penalty.[41] Not only simple folks were targeted. In
1603 the Stockholm clergyman Martinus Johannes was tortured and
executed on the grounds that he had supposedly instructed witches and
was in possession of horrible books on magic.[42] Even the most learned
men of the Realm were not above suspicion, and accusations of witchcraft could be connected to religious heresy.

Hermetic philosophers had been targeted before. Martinus Johannes's son-in-law, Johan Bure, the teacher and friend of King Gus-
tavus Adolphus, became a suspect as a result of his deep involvement
in hermetic philosophy and Rosicrucianism. Were these ideas of his in
accordance with Lutheran doctrine, and what was the difference, really,
between a philosopher and a heretic? In a chilling passage in his diary,
Bure related how he had been accused of being a heretic by Duke Carl
Philip:

> 8 [July 1621]—Saturday . . . came Duke Carl Philip and among
> other things . . . [he] asked me if I had foreseen my own time
> of death. . . . said he, Buree, You will not go to hell more than
> once. I answered, he who builds on Christ can never fall. He
> said, so says all heretics. . . . He asked if I believed in Christ, and
> if I take H[oly] communion. Yes, said I, in Uppsala and in Näs,
> and occasionally here in Stockholm.[43]

The duke's question whether Bure had foreseen the time of his death
implies that he was no ordinary heretic but also a warlock who had
been granted this secret knowledge by the Devil himself.[44] The danger of
becoming associated with witchcraft and heresy was not lost on Hiärne.
In a 1709 text, he defended Paracelsus against allegations of charlatanry
and heresy, saying that he was in almost all ways consistent with Lutheran teachings. Paracelsus had never invoked the Devil and had ad-
vised that warlocks and witches should be executed. Hiärne even commented on Naudé's exhortations against Paracelsus as an arch-heretic,

FIGURE 2. Seventeenth-century Protestants' ready association of heresy with witchcraft and Catholicism is illustrated in this piece of propaganda from the Thirty Years' War. Heresy, personified as a witch, is riding to do battle with the Catholic League and General Tilly together with an oriental warrior symbolizing tyranny. They are surrounded by Jesuits and overseen and urged on by devils. Detail of a Strasbourg print probably dating from 1631–32. Reproduction by the National Library of Sweden, call number KoB Sv. HP G.II A A.79.

stating, "Naudé, who is a Frenchman, hates Paracelsus, who never was a good Frenchman."[45]

Hiärne, in his statement to the commission on witchcraft of 1676, had denied that the Devil interacted with witches. Hence his endorsement of Paracelsus's heavy-handed treatment of witches may seem strange. It may of course be that he had changed his mind between 1676 and 1709. More intriguing, he may have had an ulterior agenda. Although mutually inconsistent, both statements serve well to disassociate

Paracelsianism from the judicial, moral, and religious realms. We must remember that the same year as his participation in the 1676 commission, Hiärne let himself be guided to his choice of spouse by a precognitive dream. It would have been inopportune for Hiärne to talk about such subtle matters as precognitive dreams and little spirit-men in the context of the commission on witchcraft.[46] In 1709, however, Hiärne publicly stated his faith in knowledge transmitted through dreams.[47] This view was completely consistent with Paracelsian writings.[48] We may safely assume that Hiärne did not believe that his dreams were of diabolical origin, and whether or not he withheld some of his views in his 1676 statement, we may surmise that he would have agreed with Naudé and other theorists of magic, that it was necessary to separate the realm of natural magic from the moral realm. The Devil was not involved in witchcraft, as he did not do his work through the world of matter but mostly through the corruption of the human mind. In this important sense, all forms of knowledge were natural to Hiärne and constituted permissible fields of inquiry and activity. The real witches, insofar as they existed at all, should be burnt, but they had nothing to do with Hiärne and his philosophy.

Hiärne brought his philosophy to the witch trials, but he had to conceal, or at least stay quiet about, many of his beliefs and positions. It is now time to go deeper. Witchcraft trials can give us an insight into common beliefs about the mutability of reality that were shared by many early modern Europeans. But the confrontational and highly charged courtroom setting can only present us with a limited understanding of early modern systems of belief, and of curious natural philosophy. It is now time to leave the legal realm, and that of human meddling in magic, to turn our attention to nonhumans. The creatures that we will discuss are the keeper entities, or trolls, who were deemed to control unmined mineral resources, and who therefore were considered to hold a significant degree of influence over the early modern mining business.

Between True Spirits and Men

Most early modern Europeans acknowledged the existence of many different kinds of beings endowed with reason, apart from men. Angels, demons, trolls, ghosts, and various other creatures populated a region in between the purely spiritual realm of God and the material world. Angels and the Devil were mentioned in the Bible, hence, they were a natural part of theology and held a strong presence in the minds of the pious.[49] But whether the other aforementioned beings were angelic, demonic, or something in between was a matter of preference and dispute.[50]

An important part of Paracelsian natural philosophy was the significance attached to subtle and spiritual beings and forces. In this area, Paracelsianism overlapped and complemented folk beliefs in Scandinavia. Hiärne stated that it "seldom meant good things" when ghosts, water keepers, elves, and similar creatures showed themselves to men.[51] His views corresponded well to those of Paracelsus, who claimed that a host of monstrous beings, such as sirens, giants, and dwarves, existed mainly as omens, to inform humankind that hard times were approaching. Ultimately, according to Paracelsus, these creatures were sent by God to warn humanity that it was digressing from the right path. But the signs were difficult to interpret as they were unpredictable and constantly shifting.[52] Hiärne seems to have agreed in that he emphasized that it was necessary to be critical and cautious when one interpreted omens, portents, and other communications.[53] Note, however, that although Hiärne was in agreement with Paracelsus, his selection of entities differed somewhat. Hiärne's water keepers and elves were typical Scandinavian creatures, whereas Paracelsus's creatures emerged from a Germanic context. Both regions, however, had a fair share of apparently sentient and thinking nonhumans living underground. As these beings often were conceived as helping or hindering miners, they were of particular interest to natural philosophers with an interest in the mining business.

Olaus Magnus, in his widely disseminated history of the Nordic peoples, *Historia de Gentibus Septentrionalibus* (1555), had stated that "it is a known matter of fact, that . . . the inhabitants of the north receive great favors and help from the trolls." Trolls belonged to the underground world of caves and mines. They took active part in the laborious process of excavating mines and sometimes appeared before miners as shadowy shapes, but their helpfulness was deceptive. They also caused pitfalls, destroyed ladders, and made all kinds of trouble, even causing regular disasters. Their ultimate aim was to kill the workers, or make them say blasphemous things, which would cause them to fall into the service of these devilish beings.[54] Magnus described Sweden as infested by trolls, devils, evil spirits, witches, and warlocks. As the last Catholic archbishop of Sweden, he wrote his book from Rome. He had been forced into exile after the Reformation initiated by King Gustav Vasa. In Magnus's view, the servants of the Devil not only dwelled in caverns and mines in the Swedish countryside, but they also held court at the royal castle in Stockholm. Miners and mining administrators all over northern and central Europe held that various nonhuman sentient beings enjoyed a heavy influence over the mining business. Magnus claimed that there were six different kinds of trolls in mines that contained metallic ore. Some were harmless while others where highly dangerous: they could kill large numbers of miners and force otherwise prosperous mining enterprises out of operation. Magnus based these views on one Münster, who in turn relied on Georgius Agricola as his source.[55]

Agricola had written at length about these denizens of the underworld in his *De Animantibus Subterraneis* (Of the living creatures under ground, 1549). This text would become well known since it was usually published together with his *De Re Metallica* (1556),[56] an influential book known and read by many early modern scholars and mining administrators. *De Animantibus* was primarily a descriptive work in the Aristotelian tradition and heavily dependent on classical authors as sources. It contained for the most part descriptions of rather mundane animals, although there were also descriptions of basilisks and dragons. The final pages of the book, however, contained descriptions of a class of creatures that did not come out of the pages of Aristotle: the demons of the mines.[57] Agricola distinguished between two major types of entities, those that caused disturbance and were dangerous, and those that were harmless or even helpful. The former type was horrible to look at and caused trouble in the mines, sometimes even killing workers.[58]

Agricola also acknowledged the existence of three types of benign or, at the very least, apparently harmless beings. The first were the cobalos or kobolder. These inhabited mines. They were about two feet high, looked and acted like old miners, and were clad in miner's garb. Their chief characteristic was that they pretended to do the work of miners although they really did nothing. Sometimes they also pelted the miners with pebbles. The second type was called *Guttell*. They lived close to humans, and performed some of the household chores in their homes, although they were rarely seen. The third kind were the *Trulle* or trolls; they were uncommon in Germany but well known among other peoples. These beings could be male or female and were known to enter into the service of men. This was especially common among the Swedes.[59]

Agricola's discussion of these beings tended toward the descriptive. There is evidence that he regarded them as semispiritual entities inhabiting a realm in between ordinary nature and the properly supernatural spheres of God and his angels. The dangerous first class of beings were described as taking "shapes" and as more dangerous than other demons, because they were more embodied in the material world than others of their kind.[60] Despite such hints, however, Agricola did not go into detail concerning these beings' demonological or epistemological status.

A man who tackled the problem head on was the foremost philosopher of the hidden denizens of the world: Paracelsus. According to Paracelsus, beings such as these were not evil. They were ghost or spirit people (*geistmenschen*), or to put it differently, just another class of people created by God to share this world with humans. They were connected to the four elements of classic antiquity: water people, mountain people, fire people, and wind people. They were also known as nymphs (or undines), sylphs (or gnomes), salamanders (or vulcani), and silvani (or sylvestres).[61] These creatures were different from humans in many ways. The most important difference, from an early modern point of view, was that they did not possess souls. Therefore they would not rise again on the Day of Judgment. Just like animals, they simply disappeared at death. All of these beings, Paracelsus explained, had an intermediate position between true spirits and men. They ate, drank, had sex, married, and begot children like humans. They were made from a much more subtle matter, however, and they had a different sense of time. They could perceive simultaneously events that happened in the future, at the present, and in the past. They were also much faster in their movements. Furthermore, each of them belonged to a specific element. The beings that could be encountered in mines, the mountain people, moved through earth and rock as humans moved through air. From this followed that they could see through earth, as humans can see through air, and that for them the sun shone through earth as it shines through our air for us. Therefore they could also see the sun, moon, and stars, just like human beings.[62]

These theories about the various elemental creatures, outlandish as they may seem, were an integral part of the Paracelsian teachings. There was, for example, a paragraph on elemental creatures in Oswald Croll's widely circulated *Basilica Chymica*, even though that work was mainly preoccupied with medicine and pharmaceutical preparations.[63]

Influential sixteenth-century authors such as Magnus, Paracelsus, and Agricola acknowledged that several different species of trolls, kobolder, and devils inhabited the underground world. By describing these entities as eating, drinking, child-rearing beings, Paracelsus placed them firmly in the realm of nature and gave them a prominent place in his philosophical system and worldview. Not all Europeans would have integrated their beliefs in these beings into a systematic and all-encompassing worldview, such as that of Paracelsianism. Even fewer would have held such systematic and positive views of the people living under the mountains as Paracelsus and his followers. In Sweden, all deliberate contacts with such entities were against the law.[64] Nevertheless, belief in beings such as these permeated all levels of society at least until the middle of the eighteenth century. For Scandinavian *curiosi*, who never were far from deep and sparsely inhabited forests, troll lore had special meaning. In 1623, Johan Bure penned an eyewitness account of an encounter with a troll said to have happened in 1598. The troll had been struck by a thunderbolt near the town of Västerås, and lay injured and motionless for fourteen days, calling to the humans passing by:

> [It] was clad in old cloth, a dirty tunic, [with] wide arms . . . had no beard, but looked like a man with coarse skin, had teeth like a boar and tusks like a boar of 2 years, they were nut-black and stood up[. The] nails [were] like a human's but long like claws[. H]e asked the people who traveled before him to turn him around, [saying that] when I am turned around on the hidden (right) side my companions will know where I am, and will take me, and he promised silver treasure for the trouble.[65]

When it was turned over on its other side by some human travelers, it immediately disappeared. The passage is highly interesting. First it makes use of what we can call credibility markers, locating the described event precisely in time and space, and giving the names of three witnesses, signaling clearly that this is a story the listener should believe in.[66] Furthermore, the story marks the troll as alien and strange. It evoked fear. Its coarse skin, overlarge discolored teeth, claws, and magical disappearance marked it as something clearly not human, and as a dangerous being who was neither an ordinary animal nor a part of the Christian community.[67] There were also many similarities between the troll and the men who described it. The story implied that this troll had been misplaced out of a cultural and social context similar to that of humans.[68] The troll wore clothes of a specific cut. He had companions who cared about him enough to seek to locate him and take him home. Furthermore, the story indicated that there was a high degree of communication possible between trolls and men. The troll had a good command of Swedish, as well as access to a silver treasure, which he could use to negotiate a deal with the humans passing by. Finally, and most importantly, the men could touch his body in order to turn him over and help him get back where he belonged. While the shapes of those who were abducted by witches could have traveled to the Sabbath in the spirit, or in their imagination, this troll was clearly nothing more and nothing less than a material being. Although a creature of magic, it was beyond doubt also a part of this world of matter. Bure's belief in the material solidity of the bodies of trolls was also emphasized in another of his annotations, where he told of how he collected the blood and bones of two trolls who had been killed by lightning in the village of Danmark south of Uppsala.[69]

The materiality of the troll is also emphasized in another way in Bure's story, through its promise of silver. Trolls, mine keepers, and the Paracelsian mountain-people controlled metallic wealth. According to Paracelsus, the people under the mountains were God's chosen guardians of nature and objects. They were assigned the task of guarding the minerals and other treasures of the earth, and to see to it that humans would not discover them until the divinely appointed moment. Although Paracelsus cautioned that these spirits could become possessed or enslaved by the Devil, they were nevertheless fulfilling a part in God's Grand Plan for humanity. To interact with these spirits was therefore something essentially good. Since they had the useful ability to predict the future, it could also be highly beneficial to listen to them when they issued warnings, or showed the way to a traveler. They could also show themselves and offer money as bribes to get rid of those who disturbed them.[70]

Directly parallel to the Paracelsian view was the common premodern and early modern belief that all mineral deposits were owned by keeper entities. In Scandinavia it was the mine keepers, or lords and ladies of the mountain, who decided where and when humans would be allowed to mine. These beings were usually solitary female entities who could interact directly with humans when they so wished. They looked and dressed like human persons of standing but had, or were assumed to have, tails like animals.[71] In some mines the keepers looked like very small old men. This may have been an influence from the lore of immigrant German mine workers.[72]

Mine keepers were invisible most of the time, but they could also shape-shift into animals. Sometimes they made their presence known as a strong force, or as ghost shapes or strange sounds. When they showed themselves in human guise, they tended to disappear suddenly without a trace. But they could also be touched physically.[73] Generally speaking, early modern Scandinavians assumed that keepers of the mountains, forests, and waters were physical beings that could touch and be touched. As attested by a fair number of court cases, sexual intercourse with these beings was not considered uncommon. It was an act that carried the death penalty.[74]

These beings had several different methods at their disposal to prevent human entrepreneurs from finding their way to the mineral veins and treasures troves. In Scandinavian lore, a common method was to bewitch the sight of the human transgressor by changing reality, or human perception of it. There are many tales in which a human discoverer of a hidden treasure or mineral vein leaves the site to return later, only to find his sight obscured or reality changed in such a way as to make it impossible for him reclaim the thing. A mineral vein could be made to disappear, leaving a previously rich mine barren.[75] Similar magic was used also by forest keepers seeking to make men lose their way in the woods, and by cunning folk of the human species, who had gained the ability to make reality appear as something other than it was.

Like these beings, central European entities were mostly invisible but could show themselves to men at will. They also had their own cunning methods to hide minerals. Either they spread poisonous fumes in the mines, or they stole away the ore itself and replaced it with worthless substitutes. These substitutes, or enchanted metals, were given the names of their destroyers. Hence cobalt was a worthless mineral that received its name from the cobalos or kobolder. Nickel, similarly, derived its name from Nikolaus, signifying a gnome or devil (cf. "old Nick").[76] These beings' replacement of valuable minerals with worthless substitutes can be seen as a variation of the Scandinavian theme that keeper entities could obscure the sight of humans, as narrated earlier. They could also make sense from the point of view of Paracelsian and Aristotelian matter theory. According to the latter, pure metals were the end product of a process of transmutation that took place under ground, through which the metals were composed from exhalations of sulfur and mercury. (In the Paracelsian view, a similar process happened, but it involved a third component, salt.) The multiple explanations permitted natural philosophers to either switch between them or stick with a single one of them. In Paracelsus's writings the two explanations of keepers and underground transmutation were combined, casting the mountain people as invisible farmers of minerals, with the capacity of destroying them before they could be gathered by humans. Worthless minerals such as cobalt and nickel ore were not species of minerals in their own right but half-baked products of a failed transmutation process that had been stopped halfway due to intervention by the people under the mountains.

Consequently, those who engaged in learned discourse of the sixteenth and seventeenth centuries had access to multiple and overlapping explanations of the way metals were generated in the earth. One of these explanations was biological or perhaps spiritual insofar as it depended on the explanatory framework of mine keepers; the other was chymical and depended on the framework provided by transmutation.

Hiärne discussed his views on keeper entities/subterranean trolls in *Den korta anledningen . . . beswarad*, a 1702 collection of letters communicating strange and curious geographical features and natural phenomena.[77] While professing a deep skepticism, Hiärne more or less passed on the question of whether such beings existed. He was, however, clear that common people were too superstitious. Granted, God could communicate warnings about coming wars and evil times to humankind through natural phenomena, but these warnings did not concern mining matters.[78] Hiärne cautioned his readers not to let themselves be influenced by superstitious beliefs, or by alleged keeper entities. There was no reason for fear, regardless of whether such entities were manifestations of the Devil, or of other kinds of spirits, "as some believed."[79] This was because Heaven and Earth belonged to the Lord, and he had given the Earth to humankind. Everything on Earth had been placed there for the benefit of mankind and was for anyone to take, provided it did not already have a human owner and that the state received its share. Consequently, Satan and his host had no influence over nature. They were not permitted to harm anyone who sought to exploit its riches. It was bad, Hiärne stated, that many were so scared. He had often heard people say that they knew where one could find a rich deposit of silver, copper, or iron, "but they do not dare to say [where] for if mine-keepers or other such troll[s] would hear them, [they believe that] their lives would not be safe."[80] Similarly, strange waterwhirls and other phenomena of such nature were not caused by lake keepers or lake trolls but had natural causes.[81]

Instead of jumping at the first superstitious explanation available, people should ask the advice of those who were knowledgeable about nature. Hiärne himself was of course such a person. Just like many other natural philosophers of the time, he sought to claim knowledge of nature as a field that was neutral in regard to religious matters. Simultaneously, explanations of natural phenomena were clearly formulated in opposition to the views held by the allegedly superstitious. Hiärne referred to the Stockholm witch trials to make his point. If physicians (i.e., Hiärne himself) had been brought in to pass their judgments at an earlier stage of the witch trials, the lives of hundreds of innocents would have been saved, who now had been killed because of the lack of knowledge among the judges.[82] As we have seen, Hiärne's role in the trials had not been as prominent as he would have it almost thirty years later. Subsequent generations would nevertheless pick up on his rhetoric, casting him as a lone voice of reason and as the main hero in the ending the Swedish witch crazes.[83]

Hiärne stayed with chymical and causal explanations when publishing on subterranean phenomena. This mode of reasoning basically proceeded from a Lutheran viewpoint. There was no need, or theological motivation, for the existence of multiple orders of intermediate beings between God and his congregation. On this central issue, then, Hiärne's views differed from those of his hero Paracelsus, who after all had remained a Catholic. Hiärne was also strongly opposed to attempts to explain regularly occurring natural phenomena by means of extraordinary causes. On this point he argued aggressively both with his contemporaries and with older authors. We can also see this line of reasoning at the witch trials, when he dismissed "lazy physicians . . . who when they cannot manage themselves in one or another natural operation . . . immediately escape to [explaining things through] occult qualities, or . . . devilish art, or magic." [84]

In an article on Lake Vättern, published in *Philosophical Transactions* for the years 1704–5, Hiärne's whole point was to present natural explanations of phenomena that previously had been thought to have been caused by lake keepers or similar creatures.[85] The Swedish Lake Vättern was known to an international readership through Magnus's *Historia de Gentibus Septentrionalibus*, in which Magnus had described Vättern as a strange lake with several marvelous and amazing properties.[86] The main issue for Hiärne was to explain the mighty waves that sometimes appeared on the surface of the lake even when the wind was calm, as well as the swift and unpredictable breakup of the ice in springtime. According to Hiärne, both phenomena had the same cause, that is, streams of air that flowed through underground channels to emerge from beneath the water.[87] Another strange phenomenon, the appearance of wandering flames on the lake and its shores, was explained by exhalation of vapors from sulfurous minerals beneath the ground, which also generated certain minerals that could be found around the lake.[88] Here, Hiärne used explicitly chymical descriptions to explain phenomena that could also have been interpreted as caused by keeper entities. Magnus had hinted that the ice breakup protected the inhabitants of the region from invading armies, but Hiärne assigned no credibility to Magnus's stories. He also made short work of Magnus's statement that, since ancient times, there had lived a wizard fettered in a cave underneath an island in the lake. He concluded his paper with the remark that a large part of what Magnus had written was "pure imagination."[89] But Hiärne would not be Hiärne if he had not left some room for otherworldly forces to operate. Thus, when he discussed the unfathomable depth of the lake, he relayed that there was a burgher in the town of Vadstena who had attempted to measure the depth of Vättern. To do so, the burgher had used a line of several hundred fathom's length, and an ax as a sinker. The line, however, did not reach the bottom. When the man, whose name was Bengt Ambjörnsson, retrieved the line, the ax was gone, but in its place was the securely fastened cranium of a horse.[90] Hiärne left no explanation as to how this had happened, just as in his *Den Korta Anledningen . . . Beswarad* he did not dismiss the possibility that there was some unknown denizen of the depth. Perhaps it had replaced the ax with a skull as a warning against this unwelcome intrusion of human curiosity into its domain? But even if this had happened, there was no reason for fear. Hiärne's world was first and foremost God's Creation. It was governed by his laws, but it was also a mysterious place, populated by angels, spirits, and maybe also trolls, all of whom fulfilled their roles in God's Grand Plan. The Devil and his witches had a place in this world, but they did not take center stage. Hiärne was a Paracelsian natural philosopher, not a demonologist. Regularly occurring natural phenomena should be explained through natural causes. It was these views that allowed him to contribute to non-Paracelsian natural philosophy and to debunk old myths about Vättern in the *Transactions of the Royal Society*.

: : :

Learned worldviews functioned well in conjunction with folk beliefs in beings such as trolls, and keeper entities associated with forests, lakes, streams, and mountains. Witches as well as trolls were vital parts of widely held cultural belief structures, according to which the material world was closely intertwined with, indeed inseparable from, spiritual and subtle realms populated by mostly unseen denizens.

Hiärne's interest in these realms was an integral part of his investigations into nature. They formed a part of his worldview, and of his philosophy of nature. To his many titles—those of chymist, royal physician, and vice president of the Bureau of Mines—we may also add that of Paracelsian natural philosopher. When Hiärne showed an interest in curious phenomena, theorized about them, and collected and systematized information concerning them, he did so as a representative of a wider philosophical current. Like many of his English and continental contemporaries, he collected evidence and anecdotes and debated and discussed them in several publications. He also planned to publish on the subject matter, his *hyperphysiologia*, as he called it. But as with so many of the works that he planned to write, he never got around to it.[91]

Many late seventeenth-century philosophers were concerned not only with the gathering of knowledge about nature but also with the banishment of moral philosophy from descriptions of the external world, and its replacement with the epistemological question of what the world is. In a sense, we can describe this as a movement away from a moral and religious discourse of nature, toward a discourse more concerned with physical objects or things. With the exception of Magnus, all of the scholars discussed—Hiärne, Agricola, Naudé, Paracelsus— shared an interest in the sorting and explaining of natural phenomena. Yet to express the issue as a matter of opposition between science and religion is to miss the point. Rather, we should see it as establishing ownership and control of nature. Like most of his contemporaries who were natural philosophers, Hiärne experienced nature as imbued with deep religious meaning: God had ordained that nature belonged to mankind. There was no need to consider the possible influence of beings such as keepers and trolls, at least not when explaining natural phenomena and when engaging in the exploitation of nature through mining. As we have seen, Hiärne was of a different opinion when discussing times and omens in history and individual destinies. In such cases there were a number of beings whose function had been ordained by God. Hiärne's sorting and discussion of natural phenomena aimed to remove one form of religious morality from nature and to replace it with another form of religious morality, more suited for the exploitation of nature's riches. In this transition, some of the phenomena and beings that previous generations had held to be part of a necessary framework in interactions with nature came to be deemed inconsequential, even if Hiärne never denied their existence per se.

It was only later that European culture in general shifted even more in its perceptions, toward an almost complete marginalization of beliefs in magic, trolls, and keeper entities. Seventeenth-century epistemology, or knowledge, about things material was still quite fluid and open-ended. Humility, not skepticism, was the proper stance when beholding the vastness of nature's mysteries. This position would change. As I indicated in the discussion of Cartesianism, a number of late seventeenth-century thinkers began to assume a radical division, even a gulf, between the realm of spirit and the material world. Such notions caught hold strongly in the intellectual climate of England during the second half of the seventeenth century and the first decades of the eighteenth. Within this interpretative framework, access to the world of spirit, and contacts and intercourse with spirits were increasingly described as rare and singular events. They were either a sign of God's particular grace, of diabolical infestation of the gravest significance, or simply of delusion. Simultaneously, many English chymists, and Robert Boyle in particular, were highly critical of Paracelsianism.[92]

It would take some time before this discourse would have an impact in Europe's Protestant north. The Paracelsian framework made a lot of sense from the point of view of seventeenth- and eighteenth-century Scandinavians. Much of the chymical worldview of the Paracelsians had been reinterpreted to fit snugly with Lutheran theosophy, such as that proposed in Rosicrucian circles, and indeed endorsed by Martin Luther himself.[93] Interaction with, for example, trolls and keeper entities, as well as precognitive dreams and second sight were considered common-place. Widespread popular beliefs had it that a range of invisible enti-ties lived in forests, streams, caverns, mines, and even beneath stables and farmhouses. Supernatural guidance was not uncommon. Lutheran theologians explained that disasters were signs of divine retribution and that unusual natural phenomena were divine warnings. Bishops pro-claimed that God communicated directly with humankind through signs such as thunderclaps. Hiärne's contemporary, Bishop Jesper Svedberg, held thunder and lightning to be signs that should be interpreted, and another bishop, Andreas Rhyzelius, stated in 1721 that thunder and lightning were intended as warnings from God to humankind.[94]

It is likely that this cultural context contributed to making the defi-nition of matter an important issue for the Bureau of Mines chymists. From their preoccupation with mining and smelting, it followed that they interacted professionally with miners and other rurals, and had to relate equally to popular and learned beliefs. Hence, their self-appointed task of transforming chymistry into an Enlightenment enterprise, in-cluded dealing not only with chymistry's heritage of transmutation but also with defining chemical material reality as something apart, and dif-ferent from, popular perceptions. Chymistry had a major role in discred-iting the material reality of trolls, witches, miraculous transformations, and magical causation because it was chymistry that established itself as the authority on what matter was; thus it also had the final say on what matter could *not* be and do. As I claimed in the opening chapter of this book, an important part of this process of defining matter occurred among officials attached to the Swedish Bureau of Mines. It is now time to present this environment and to delineate Hiärne's contribution to the institutionalization of chymistry at the Bureau.

3

Chymists in the Mining Business

The Meaning of Chymistry in the Seventeenth Century

To many early moderns, a true chymist or adept was a person who had achieved ultimate mastery of the material world. He had transcended the ordinary hopes and fears of mortal men. Through skillful manipulation of matter he had created the philosophers' stone—a substance that allowed him to manufacture an enormous amount of wealth (i.e., gold). Alternatively or simultaneously, he had learnt the secret of the universal elixir, which prolonged his life beyond that of ordinary men. Early modern chymistry was a dream and a vision of what a pious and dedicated artisan could achieve if he only committed himself—body and soul—to his work. This was the inherent promise that lay at the heart of what was called both the Royal Art and the Great Work.

Just like the discursive realm—and practices—of witchcraft, the chymical enterprise was completely integrated into early modern European culture. It was a part of the artisanal world of towns and cities. In a sense the adept was an idealized traveling journeyman who had learnt the secret of gold making, rather than, say, a novel method of how to dye fabrics or build a vault. There was nothing epistemologically outlandish about chymistry. That is, most people had good reasons to believe it to be possible. The promises and scope of the art were consistent with most premodern and early modern theories of how changes occurred in the material world. Similarly, many chymists regarded chrysopoeia (that is, the chymical transmutation of lesser metals into gold) as a natural process. The gold-maker simply re-created a process that already occurred in nature on an everyday basis. Accounts and stories about successful adepts and successful transmutations were heard everywhere. Some of them seemed very credible and could be connected to reliable witnesses, or to nuggets of gold said to have been manufactured through chymical transmutation.[1]

Early modern chymists did not limit themselves to attempts to transmute lesser materials into gold and the search for the universal elixir. As shown in a number of illuminating recent studies, their knowledge intersected with that of other artisans. Chymistry, as it was practiced in early modern Europe, was highly concerned with technical and medical innovation. Most chymists worth their mettle claimed to possess knowledge of a number of lesser secrets: for example, knowledge of *particulares*, that is, methods to effect simple transmutations, such as turning silver into gold.[2] Many also claimed to have discovered cures for specific diseases; others claimed knowledge of how to produce unusually colored glass, or how to make a paint that would prevent the keels of ships from being destroyed by shipworm. A special class of claims concerned ways to improve smelting practices at mines, to get a greater yield of metal (which would be manufactured through chymical augmentation). This latter type of claim was, of course, of special interest to mining officials and something we examine more closely in this chapter. These examples show that chymistry in the seventeenth century provided a theoretical framework for thought and action in a number of fields. Since there were no guilds for chymists or any other formal education to be had in the area, the field was open for all. Individuals could, to use the words of Tara Nummedal, use chymistry as a "high-risk, high-reward game." By employing the language of chymistry rather than that of, for instance, goldsmithing, a goldsmith's apprentice could become a Paracelsian doctor, and then a smelting expert. Hence the art could provide a framework that allowed the practitioner's identity to be transformed and his or her social position advanced. The practitioner could end up as a senior advisor to a prince or on the gallows, or both in turn. Chymistry's ambiguity and fluidity made it an excellent vehicle for ambitious individuals bent on bettering their lot.[3]

To European rulers, chymists possessed a type of knowledge to which it was desirable to have access. Although outlooks had begun to change already in the previous century, seventeenth-century European elites showed an ever-increasing interest in fiscal matters and natural philosophy. The common eighteenth-century trope that knowledge should be useful for the public was also beginning to take hold—simultaneously breeding a disdain for less worldly and more contemplative pursuits. Growing state bureaucracies, debts incurred through an ever-increasing need to publicly manifest wealth and power, and heavy military expenses also forced many rulers into a search for new sources of wealth. Chymists and other project-makers promised quick and easy solutions to these problems, and especially in German-speaking central Europe, chymistry managed to secure an unparalleled place at the center of cultural life and economic discourse.[4]

Whether one sought to develop a new product, solve a specific technical problem, develop industry and manufacture in one's realm, or just felt needy of huge amounts of chymically manufactured gold, chymists offered advanced technical, chemical, and economical knowledge on how to do it. Chymists worked on the assaying and smelting of metals to improve the mining business. They were engaged in pharmaceutical development of new medicines. They could be made responsible for the large-scale manufacture of goods in factory-like settings. And there was, of course, always that elusive possibility that the chymist literally would create gold and make all other types of productive activity unnecessary for his or her patron. Even in the many, many cases when no gains came out of the chymical experiments, the patronage of chymistry nevertheless reflected well on the monarch, prince, or nobleman who sponsored it. It enhanced personal reputation and glorified the realm. Even when chymists failed to live up to their promises, chymistry nevertheless worked well as an effective form of conspicuous consumption. Hence, seventeenth-century chymistry cannot be understood outside of the economic contexts in which it thrived. But it could also be used to support other agendas, whether they were political, religious, learned, or the pursuit of status and glory.[5]

It is important to remember that this was *real* knowledge. Many chymists spent their lives investigating the transformative processes of nature and industry. We cannot say whether most of their efforts to produce improvements and useful discoveries were in vain, but the crowning achievement of this form of economically oriented chymistry is impressive enough. It was Johann Friedrich Böttger's 1708 discovery of the recipe for true porcelain, which previously had been a well-guarded Chinese secret. After making his discovery, Böttger also set up and developed the porcelain works of Meissen, thus creating the model for all further European porcelain manufacture. We can also quote Henning Brand's discovery of phosphorus, which he made after a painstaking series of experiments in the distilling of urine.[6]

In Sweden too, chymistry was pursued as a part of the country's policies to increase revenues and to bolster its reputation. In this chapter we investigate how chymistry was transferred from a continental European context, to become institutionalized in the mining bureaucracy of the Swedish state. I argue that it was through the language and practices of chymistry that natural philosophy made its first strong inroads into the Swedish state apparatus. The main venue for these inroads was the Bureau of Mines, and it was through Urban Hiärne, the protagonist of the previous chapter, that chymistry received the attention and support of the Swedish bureaucracy.

It is necessary to understand the institutionalization of chymistry during the decades before 1700 in order to understand the later developments during the eighteenth century. There was a high degree of continuity. Later development in chemistry built on institutional foundations created for Paracelsians and others, working within a paradigm of metallic transmutation. The main point here is not to study Hiärne's Paracelsian thought or even the experiments in chrysopoeia and medicine making that he pursued while attached to the Bureau of Mines but rather chymistry's social context as a part of a bureaucracy. Arguments for its utility and attempts to make it useful in the business of mining take center stage. And I argue that it was through Hiärne's efforts that chymistry was articulated as a publicly useful and economically oriented enterprise in the Swedish context. Hence, an important point established in this chapter is that something very peculiar happened to chymistry when it was allocated a space—and a considerable budget—within a bureaucratic organization. It began, almost of its own volition, to transform itself into utilitarian eighteenth-century chemistry.

A World of Patrons and Clients: Late Seventeenth-Century Stockholm

Before the seventeenth century, the heavily forested and sparsely populated Realm of Sweden had been something of a European backwater. Then, the country's victories in the Thirty Years' War led to increases in both territorial size and influence. Stockholm changed from a medieval town to the capital of a small empire. The old town around the royal castle was transformed as room was made for offices for bureaucrats and administrators as well as a growing number of private palaces accommodating the country's nobility and major traders. The new and vastly improved bureaucracy enabled the country to take a firmer grip on its economy and territory. Stockholm became a center of control of the outlying provinces. Reforms had been instigated by the legendary chancellor of the Realm Axel Oxenstierna. The state, which previously had been run mostly by the king himself and his office of secretaries, the Chamber, was now reorganized around about a dozen bureaus (*kollegier*).[7] These functioned as government departments with wide responsibilities and carried enormous authority. Each bureau commanded a well-staffed central office, as well as various local officials. At their core was a governing board, a panel of senior officials bearing the titles of assessors and councilors. The governing boards of the bureaus had a direct link to the monarch, as their presidents were also members of the Council of the Realm.[8] The Bureau of Mines was founded in 1637 as the twelfth bureau and was the Swedish state's foremost instrument to exercise power over, and gain advantage from, the country's mining business.[9]

The Swedish creation of a professional mining and smelting administration in the 1630s was modeled on German examples. It was an instance of the application of the directorial system (Ger., *Direktionssystem*) that had been championed by, for instance, the dukes of Welfe in the Rammelsberg and Upper Harz already in the mid-sixteenth century. As metal production became an important source of income for the state, control tightened. Private ownership of mines and smelting works was permitted, but owner influence was severely limited. All major organizational, financial, and technical decisions were gathered to the state administration, which was organized according to hierarchical principles. As mining historian Christoph Bartels notes, this "typical absolutist approach to metal production . . . prevailed in many European mining districts, though not in England."[10]

These highly interventionist policies depended on the concept of the *Bergregal*, which separated ownership of the land from ownership of the mineral resources beneath the ground. The latter belonged to the sovereign (or in some cases the feudal landholder) of the realm. Hence a basically feudal legal framework made it possible for early modern states to become major players in the mining business—either as direct owners of mines and smelting works or as instigators of detailed legislation aiming to guarantee the state a substantial share of the proceedings of private mining enterprises.[11] Policies received different expressions in different contexts. In Saxony, for example, increasing numbers of smelting works were transferred to the state from the sixteenth century onward, and the remaining works that were privately operated were subject to stricter control. This centralizing tendency reached its peak at about 1710, when all of Saxon smelting became state-run.[12] Similar measures were implemented by the Swedish state. The *bergsmän* (i.e., mine shareholders cum small-scale smelters) of the great copper mine of Falun were required to deliver their product—coarse copper with about 90 percent copper content—to be refined at the state-operated works at Avesta. Other mines and works, such as the silver mine at Sala, were directly operated by the state. The backbone of the Swedish mining enterprise, the highly profitable iron industry, was to the most part privately managed by ironmasters of high social standing, whose ironworks nevertheless were subject to the tight regulation and control of the Bureau of Mines.

As mentioned in the previous chapter, Hiärne's patron Claes Tott died in 1674, and the responsibility for Hiärne was taken over by Tott's older relatives Sten Bielke and Bengt Horn. These men were heads of aristocratic families and held seats in the Council of the Realm. In 1683, Hiärne's new patrons managed to convince the king that their charge was worthy of royal patronage. But it was not Hiärne the artist, poet, and playwright who would receive the state's support. Instead, Hiärne was to become a royal chymist with a hefty salary, an assistant, and a large budget for a laboratory.

Highly sensitive to the sentiments of his times, Hiärne had already begun to abandon art, poetry, and literature for medicine and chymistry in the early 1670s. He became a chymist during his stay in Paris in 1670–72. When he moved to Stockholm in 1674, chymistry had become one of his main areas of interest and knowledge, and he would continue to pursue it for the rest of his life. In Stockholm he established a private laboratory, which he shared with some medical friends, so we must surmise that in 1683, Hiärne had already amassed a significant knowledge of chymistry.

But where would one graft a chymical laboratory onto a bureaucratic and hierarchical organization like the Swedish state? In fact, there already existed an official laboratory, attached to the government's Bureau of Mines. When the Bureau had been founded in 1637, one of its first tasks was to establish a chamber of assaying, in which the minerals of the various mining sites of the Realm would be investigated. In 1639 it also set up a laboratory and hired a German assayer, Georg Tebicht, as its leader with the title Laborant. The chamber of assaying and the laboratory shared servant staff and were placed in the Bureau's main office in the royal castle in Stockholm. When Tebicht died in 1641, his office was given to an apothecary, Liborius Finzenhagen, who also preserved his widow. Finzenhagen, however, was deemed to be a difficult person and a drunkard and was sacked in 1642. For about a decade the laboratory was closed, and then Finzenhagen was reinstated and stayed at his post until he died in 1664. Again the laboratory was left empty until a Paracelsian and chymist, Wendelinus Sybellista, secured the post as its leader. Sybellista had gained the ear of the Bureau by claiming to have knowledge of a secret process that would improve the extraction of copper from its ore. A new building was erected for him, but in the end Sybellista did not manage to secure the funding for his projects, and he was flatly rejected when he applied for money for experiments in gold making. He was forced to leave in 1668. A few years later he was succeeded by the apothecary Christian Heraeus, who seems to have done little work in the laboratory: in 1674 it was torn down to give space for the Royal Stables.[13]

There was a reason the Bureau could handle the laboratory and its leaders with such relative lack of interest. The laboratory's main task was really to function as a pharmacy. As part of the Bureau of Mines' responsibility for running the state-owned mines, it employed surgeons who cared for the workers at these mines. The laboratory was charged with providing medicines for the Bureau's surgeons. When it lay dormant, medicines were purchased from other pharmacies.[14] There had been periodic interest from the Bureau to expand the tasks of the laboratory, and both Finzenhagen and Sybellista had made suggestions that had gained a degree of support from the Bureau. It was only when Hiärne was appointed, however, that the head of the laboratory was able to secure patronage and funding for more far-reaching and extravagant projects. Although Hiärne took over a lowly position in the Bureau of Mines hierarchy, he and his patrons transformed this position and the Bureau's laboratory into a showpiece for the Swedish state. When Hiärne petitioned for state support for a laboratory, the practice of gold making was downplayed, and even the manufacturing of specific medicines and other inventions did not take center stage. Hiärne's rhetoric focused on the need to raise Sweden's status in the eyes of foreign nations. The king and the Realm would gain prestige from having their own laboratory. As Hiärne put it, "there is no King, no potentate, no Prince, yes there is not even a Count abroad who does not have a chemical laboratory, each after the proportion of his estate."[15] Sweden's new chymical laboratory was primarily there to glorify its monarch. In addition it would be preoccupied with various types of useful research to improve the country's economy. Although the laboratory retained its old functions at the Bureau of Mines, it was also reinvented along the principles of the princely laboratories that were dispersed throughout the Holy Roman Empire.[16]

Hiärne was installed as director of the laboratory in 1683. Simultaneously the president of the Bureau of Mines since 1679, Hiärne's patron Bielke, saw to it that he was given a seat as an assessor on the Bureau's governing board. This event completely changed the character of the laboratory in that Hiärne now had a voice in the leading group of the Bureau, and was able to defend his interests and affect the Bureau's general policies. The laboratory became an independent unit tied to his person. The new institution was to be well financed, with an annual budget of 3,100 Swedish silver dalers, of which 800 was to be Hiärne's salary, and 300 was allocated as salary for a skilled laboratory assistant. According to the initial plan, the money would be drawn straight from the king's personal ready money. In 1684, however, Bielke saw to it that the money would be paid instead by the Bureau of Mines.[17] This was good news for Hiärne, as the bureaucratic Bureau was a reliable employer, and he could do more or less as he pleased as long as he delivered medicines to its surgeons. A new stone building was erected for the laboratory in central Stockholm in 1685–86. Hiärne had to abandon it some years later but subsequently bought the stately Gripenhielmska Palace on the centrally placed island of Kungsholmen. It testifies to his influence at this time (1694) that the king and the Bureau of Mines received no formal prior notification but nevertheless decided to pay both the purchase sum and the hefty cost of renovation.[18] Hiärne's salary, budget, and status in the organization were elevated in comparison to that of his predecessors. But there was also continuity. The laboratory's responsibility to manufacture medicines continued for a long time and was extended to provide medicines for the armed forces as well. It is important to note that natural philosophy in any form had received little attention at the Bureau of Mines prior to the 1680s. The concept that the aristocrats themselves should get their hands dirty in the practice of mining and smelting was as yet something of a novelty. Many Swedish aristocrats of the mid-seventeenth century cared for learned pursuits, but it was law, as well as the humanities, art, and literature that took center stage. Chymistry too had had difficulty getting a foot in at the Bureau and was unpopular among certain parts of the ruling elite.[19]

Hiärne's activities were something new. Unlike previous employees of the Bureau, he was neither an artisan nor an aristocrat, but a physician and natural philosopher who had risen to his position as a client and protégé of aristocratic and royal patrons. Although Hiärne's chymistry was new in the context of the Bureau of Mines, it conformed

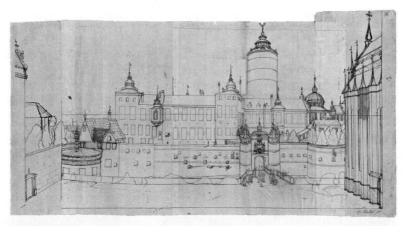

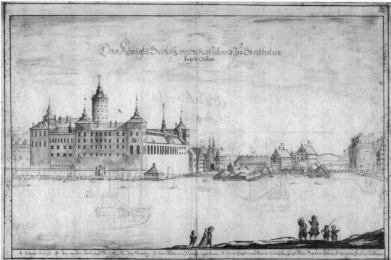

FIGURE 3 shows a sketch of Stockholm Castle (drawn by Erik Dahlberg in the beginning of the 1660s) from the west and the rotunda of the Chymical Laboratory (*middle left*).

FIGURE 4 shows a sketch of Stockholm Castle (drawn by Erik Dahlberg in the beginning of the 1660s) from the east and the island of Helgeandsholmen, with the Royal Mint located in between. Reproductions by the National Library of Sweden, call numbers KoB Dahlb. Handt. 1:21, KoB Dahlb. Handt. 1:20.

to a pattern discernible in other parts of Europe. Pamela H. Smith has shown how chymists in the seventeenth-century Holy Roman Empire functioned as go-betweens who linked the artisanal world of crafts with princely courts. Just like Hiärne, the most successful of these chymists had learnt to present and handle themselves in polite company and could understand the rules of intrigue and patronage at court, which made them socially acceptable among the elite. Simultaneously they had access to the knowledge and culture of artisans. They knew, or claimed that they knew, how to manufacture valuable objects and how to set up advanced production processes. This information enabled chymists to co-opt, as Smith puts it, "artisanal modes of seeing and doing" and make them available to territorial rulers.[20] Chymsts were part of a larger group of projectors, would-be economic advisors, and other actors around early modern states that offered rulers knowledge of how to make money through the manufacture of goods.[21] Gold making was just one of several potentially useful pursuits on offer. Wakefield expresses the plight of early modern rulers well. How should one choose advisors? "If one could train or recruit . . . less ignorant, more industrious, less haphazard, and more honest officials—or if one could find a way to discipline them, then the sovereign revenue would benefit."[22] When the Swedish state turned to Hjärne's chymical knowledge, it was as part of an attempt to discipline chymistry, and to make it reliable, trustworthy, and patriotic knowledge.

A zealous patriot, Hiärne enlisted chymistry for the state's efforts at reform. He also contributed to bureaucratization. He successfully created a place for investigations into nature and for natural philosophy, situated in the core of the state apparatus. He created an institutional environment for chymistry in the sense that he made sure chymistry was assigned a specific number of tasks within the organization, and the budget needed to carry them through. The laboratory in which he conducted his research also served a representative function glorifying the monarch and the Realm. Hiärne bolstered the status of his sphere of influence, showcasing through his publications the successful innovations that had been made under the patronage of the king. And he took the process of linking the artisanal world of crafts with princely courts one step farther. He was much more than a go-between.[23] In the context of the Bureau of Mines, he should above all be seen as an institution builder. By introducing artisanal knowledge into the state bureaucracy, he created a powerful contact zone in which artisanal knowledge could be transformed into the knowledge of high-status mining officials. This contact zone linked together noblemen and artisans, Swedes and Germans, chymistry and smelting, and it would continue to exist and expand long after Hiärne's demise.

Making Chymistry Mining Knowledge

When Hiärne became attached to the Bureau of Mines he was a forty-two-year-old poet, playwright, artist, medical doctor, natural philosopher, and aspiring chymical adept. But he knew little of mining or of administration. One of his first tasks was to further his knowledge in his new areas and to contact others who possessed the kind of knowledge that he deemed suitable. Instrumental to these ambitions was his student Erich Odhelius (1661–1704). Odhelius was a scion of one of the most learned family networks of the Swedish Realm.[24] He had begun his studies in Uppsala in 1676 and had soon gravitated toward the medical faculty, which was where natural philosophy had its strongest foothold. He boarded in the home of the professor of medicine Petrus Hoffwenius, who had, along with Olof Rudbeck the elder, been the instigator of the Cartesian battles in which Hiärne also had played a part as a student. Hence Odhelius was trained as a medical doctor by the same headstrong natural philosophers who had been Hiärne's teachers fifteen years earlier. It was not particularly strange that Odhelius's manifest interest in medicine and chymistry soon would bring him to the doorstep of Hiärne's laboratory in Stockholm. The first evidence of contact between Odhelius and Hiärne derives from February and March 1681. During this period Odhelius stayed, probably as student, in Hiärne's laboratory. It was as yet only a private laboratory established by Hiärne and some medical friends. Odhelius, however, soon became more than a student, eventually becoming Hiärne's client, friend, ally, and closest disciple.[25]

During the following years, he would travel Europe and study practical and theoretical chymistry, medicine, and pharmacology, mining crafts such as smelting and assaying, as well as administrative skills. He would visit Hungarian mining towns in the east, Naples in the south, and London in the west, where he would teach Robert Hooke some manners in polite communication and learn to not give away too much of his knowledge when prodded by the shrewd Robert Boyle.

Hiärne's chymical activities were tied to the Bureau of Mines at about the same time as Odhelius left on his first journey.[26] Through Odhelius's letters we can study Hiärne's attempts to connect his knowledge with the business and administration of mining as conducted at home, on the Continent, and in England. Odhelius would serve as Hiärne's ears, eyes, and hands abroad. He forwarded packages, letters, and manuscripts, recommended new laboratory assistants, and informed Hiärne about new curiosities and the latest gossip, both from the world of the learned and from the shady world of traveling chymical adepts.[27] He answered questions, supplied information and made new contacts.[28] Later on Odhelius would travel as an official representative of the Bureau of Mines and was given more formal tasks to collect and forward information directly to the Bureau.[29]

As mentioned earlier, physics and chymistry did not play a significant role at the Bureau before the 1680s. Odhelius's travels, and also Hiärne's, were an effective way to circulate knowledge on these matters. Knowledge exchange through journeys became an important component in the build-up of institutional structures for natural philosophy at the Bureau. Odhelius's journeys also set the pattern for coming generations of officials, many of whom would more or less follow his itinerary in their own attempts to gather information for the Swedish state's mining enterprises.[30]

Odhelius's first destination was Saxony, a thriving electorate in the eastern part of the Holy Roman Empire. Saxony was a center for pharmacology and chymistry and included important cities and towns such as Dresden, Leipzig, Meissen, and Chemnitz. Above all, however, Saxony held the town of Freiberg and its famous surrounding mining district. The town was the center for Saxon silver smelting, and an important regional and European site for the teaching of the mining and smelting arts.[31] In the late Middle Ages, Freiberg had been one of the most prosperous cities of the Holy Roman Empire. Apart from silver, Saxon copper production was also significant, although a more important copper-mining district was that of the neighboring lands of the Counts of Mansfeld, which in great part were under de-facto Saxon rule.[32]

Odhelius arrived in the Saxon capital of Leipzig in the autumn of 1683, after a pleasant and uneventful journey.[33] He was received in the home of Heinrich Linck, collector, chymist, and proprietor of the pharmacy of Zum Goldenen Löwen. He was also one of Hiärne's correspondents.[34] Odhelius was to stay with Linck to learn the laboratory arts and materia medica of the pharmacy trade. He believed it would take him approximately a year and a half. When he was finished he would continue on to Freiberg to learn mining arts and assaying. Odhelius settled in in his new home and was well content with his host. He and Linck worked together in the laboratory, and sometimes they visited Linck's patients together.[35] With an eye to his return to Sweden, he was also preoccupied with building up collections to accompany his newly won knowledge. He collected minerals and manuscripts, and created a small personal pharmacy containing medicines from both Sweden and Leipzig. He had a box made for his pharmacological materials and ordered an expensive apothecary's cabinet from Augsburg.[36]

Odhelius's journey was something in between a grand tour of Europe and a journeyman's journey. On the one hand, he was a comparably rich traveler. (Presumably, he traveled at the expense of his family.) As is apparent from his purchases, he did not plan to return to Sweden like a journeyman, with his possessions in a knapsack. He could not carry the many items he collected but had to transport them in a carriage. On the other hand, he traveled in order to learn the arts of artisans and craftsmen. He was as yet not a nobleman, and he did not have a manservant.

In keeping with his plan, Odhelius focused primarily on learning the arts of the apothecary. As yet he had no knowledge of assaying, which was a completely different type of knowledge. The apothecary learnt to recognize and prescribe the many animal and herbal drugs of Galenic medicine, and how to chemically decompose substances into their consistent parts, and to recombine them in order to prepare medicines.[37] The assayer's knowledge concerned minerals and precious metals. He learnt how to analyze stones and earth samples for their content of valuable minerals, and he knew also how to ascertain the true metallic composition of objects made of, or made to look like, gold, silver, and other metals.[38]

Although assaying and pharmacology were two distinctly separate areas of knowledge, a single individual could learn them. Odhelius, Hiärne, and other chymists considered them both parts of a more generic chymical knowledge. When Odhelius recommended one of Linck's apprentices as a laboratory assistant to Hiärne's Stockholm laboratory, he explicitly commented that the man had a sound knowledge of the pharmacopoeia of Galenic medicine and knew how to prepare chymical medicines ("in Officinalibus Galenicis & Chymicis praeparandis"). The man, however, knew nothing about assaying. If Hiärne wanted a more skilled assistant, he would have to find someone who had more experience.[39] Apparently it was an assayer that Hiärne required, for the young man does not seem to have been sent to Stockholm.

Hiärne, as a renowned doctor of medicine, had connections among learned men of letters, medical doctors, and apothecaries, but he seems to have had few or no European contacts with access to, and knowledge about, the mining business. One of Odhelius's tasks was to make contact with people who had a knowledge of mining. The problem was that Odhelius, being himself a medical student, had little knowledge in the area. To present himself in the mining districts of Saxony, Odhelius first had to learn more about mining and assaying. In a number of letters to Hiärne, Odhelius excused himself, saying he was not yet ready to go to Freiberg. It seems as if he could not find a proper way to introduce himself, as he did not know anyone in the mountain towns. He explained to Hiärne that he did not want to contact anyone in Freiberg until he had gained a more thorough knowledge of mining matters. He did not want to go until he had visited the mountain towns together with Linck. He also did not know where to find a good teacher, or in which town such a teacher could be found.[40] All in all he seems to have been uncertain of whether he would be received and whether he knew enough not to make a fool of himself.

In June 1684 Odhelius finally traveled to the mountain towns together with Linck and a few other companions. The leader of the troupe was one Doctor Helzel, whom Odhelius described as having excellent knowledge of mining. Helzel had the necessary contacts and friends that Odhelius lacked, and through him "everything stood open for us to see, and I went down into every mine that we could get to." Throughout this journey, Odhelius made many friends of his own in the mining towns, but he did not visit Freiberg.[41]

The trip to the mountains gave Odhelius many of the contacts he needed and gave him a new stimulus to improve his knowledge of the mining arts. He decided to enroll for a course in assaying held by Doctor Helzel.[42] At this point, his knowledge seems to have been diminutive. The only book he had bought was Georgius Agricola's *De Re Metallica*. He had also composed a list of suggested reading for Hiärne's perusal, based on what was available among Leipzig's booksellers. There was "Leenheusen," by which he probably meant Georg Engelhard von Löhneysen's *Bericht vom Bergwerk* (1617). Furthermore "Faxen," in all likelihood Modestinus Fachs's *Probier-Buchlein*, as well as "Mathesii Postilla," i.e., Johann Mathesius's *Bergpostilla* (1578).[43] It was not a bad selection of books to begin with, but most of them were dated. Agricola's book was not so much a handbook for miners as an overview of the area for the instruction of the interested Renaissance public. The last known early modern edition was published in 1657. Of the books on Odhelius's shopping list, Löhneysen's was the only one that had been written in the seventeenth century.[44] On the other hand, there would not have been many other books available in 1690. For example, Balthazar Rössler's *Speculum Metallurgiae Politissimum* (published in 1700) circulated as yet only in manuscript copies, and Abraham von Schönberg's *Ausfürliche Berg-Information* would not appear until 1693.[45]

In springtime 1685 Odhelius finally arrived in Freiberg. The town had as yet no formal institution for the teaching of assaying. Prospective students had to seek out their teacher themselves, and the cost was higher than that of learning any other craft.[46] Odhelius made his agreement with Herr G. Süssmilch, who pledged to lodge him, teach him assaying, and give him a testimonial of his apprenticeship when he had completed his studies. Odhelius established himself quickly. In December he had become friendly with the town's leading mining offi-

cial, Abraham von Schönberg (Oberberghauptmann in Freiberg, 1676–
1711), who permitted Odhelius to accompany him on an inspection
tour.[47] Schönberg would have acknowledged his ambition to study the
mining arts. In 1702, he suggested that a special fund would be created
to which promising Freiberg students could apply for stipends to further
their knowledge of land and mine surveying and assaying.[48] The Bureau
of Mines' sponsored travelers may have inspired him.[49]

Since 1644, the Bureau employed two salaried auscultators, who
were a kind of trainee for the positions of assessor. In 1693 the office of
auscultator was turned into an unpaid position at the lowest level of the
Bureau hierarchy. The two salaries were turned into stipends that the
now much more numerous auscultators could apply for. In 1699 two
new stipends were created. From then on there existed four stipends at
the Bureau. The two older ones came to be tied to chymistry, and the
two new ones were explicitly tied to studies in mechanics.[50] The mining
authorities of Joachimsthal (1713) and Chemnitz (1735) would follow
suit and enact similar stipends within a couple of decades.[51]

Odhelius also learned land and mine surveying in Freiberg, and continued to add to his collection. He gathered manuscripts and ordered
mining and smelting tools and equipment for assaying. By now, Odhelius had gained enough knowledge to begin to think about whether he
would become a medical man or continue to pursue mining knowledge.
As he put it (although he wrote in a characteristic mix of Swedish, Latin,
and German), "both professions lay claim to the entire man and are
extensive enough, [at least] if one wants to achieve something well in
[either of] them." Eventually he decided that although he would like to
become an assayer, he would at least finish his studies in medicine so
as to not let his previous studies come to nothing. He also sketched out
where he would go to further his knowledge of mining: he would travel
to Böhmen, Tyrol, Eisleben, and Goslar and make his return to Sweden
from Norway.[52]

In 1683 Odhelius had come to Saxony to learn the skills and arts
of the apothecary in the shop of Herr Linck and those of the assayer in
a suitable but unspecified location. Two and a half years later he had
achieved a working knowledge of these crafts. But it was not the intention of his patron Hiärne that he would become an apothecary or an
assayer. Odhelius was destined for a prominent position at the Bureau
of Mines. His new skill in the two principal early modern crafts that
manipulated and wrought change on matter was a good foundation.
From Hiärne's point of view, however, they were no more than tools
that would permit his client and charge to reform Swedish mining and

smelting from within. For Odhelius to have a complete education, it was also necessary that his knowledge be situated within a theoretical framework that encouraged innovation and discovery. This framework was provided by chymistry.

Meeting Chymists

Chymistry was a strong presence in Odhelius's letters. He continuously informed Hiärne of rumors concerning adepts. Linck, Odhelius's host in Leipzig, conducted chymical experiments that initially he kept secret; eventually, however, Odhelius was invited to participate. During his first time in Leipzig, Odhelius also spent much time in the company of one Meisner, a student of medicine from Wittenberg who professed to have devoted his life to the search for the philosophers' stone.[53] About a year later Odhelius was instead enjoying the company of one *Magister* Rothmaler, who had become a friend of both Linck and Odhelius. Rothmaler, claimed Odhelius, had a *particulare* and said he was on his way to discovering a *universale*.[54] Robert Boyle may also have shared Linck and Odhelius's high esteem of Rothmaler, who is likely to be Erasmus Rothmaler, from whom Boyle received an inscribed manuscript that can still be found in the Boyle papers.[55] If the identification is correct, it gives us a tantalizing glimpse of deeper interactions and exchanges that may have existed among the transmutative chymists of the Holy Roman Empire, England, and Sweden.

Beginning with his very first letter from abroad, Odhelius dutifully reported chymical rumors and gossip as well as his own endeavors. His first letter shows his inexperience of the world of chymistry in an almost touching way: he wrote that he had heard from a French chef in Ultenaw that there were three reputable and God-fearing adepts in Hamburg, and that he had asked a man to investigate the matter. The quality of Odhelius's information soon improved, and simultaneously his skepticism increased. As Linck was well known, various shady individuals who claimed to be adepts often turned up on his doorstep. Odhelius's skepticism, however, did not as yet concern the chrysopoetic project as a whole. Rather it indicated that he had gained in experience and had learnt to use the contemporary discourse on how to recognize false adepts, or "Pseudo-Adepti."[56] Morality, knowledge, and social position were inextricably linked in such discourse. As many other chymists did, Hiärne and Odhelius held that completion of the Great Work depended on God's grace. There was a miraculous component to the work, and no matter how skilled the operator, the simple manipulation of the material world was not enough.[57] As it was unlikely that anyone except the most noble and pious would be granted these exceptional favors from God, the morals and perceived personal qualities of a true adept became an issue. Many of Odhelius's letters were concerned with the exploits of one Helbig, in all likelihood Johann Otto Hellwig, who seems to have been the most visible chymist in the Holy Roman Empire at this time. Odhelius was skeptical of whether Helbig could be a true adept, as he did not like the man's high profile, antics, and bragging.[58] In another passage, he doubted the knowledge of one Doctor Köser, who had offered to show Linck the chymical method leading up to the famous secret process of the red lion (*usque ad Leonem rubeum*) in exchange for money, yet he did not even know how to cure a fever.[59] It was also necessary to safeguard against frauds, as when an Italian "tramp" visited Linck "and made a terrible noise about his chemical operations," only to be shown the door while screaming curses at Odhelius.[60]

Odhelius's meetings with chymists continued in Freiberg, but the many meetings only served to increase his skepticism. "By the way, I have met Alchymists this week, who soon will poison all of Europe, and in particular Freiberg will soon be completely turned into gold."[61] Odhelius had met the elderly Johann Daniel Kraft, the previous director of manufactures in Dresden, and one Starcke, who had arrived with Rothmaler, and who had traveled extensively through Europe. "Although no art, which now in Europe is considered rare, is unknown to him, he admitted that he knew nothing of importance; since he maintains that nothing is fit of his admiration except Gold-making." With Starcke was also one Marechalli, and the two men were apparently to conduct chymical work together in Freiberg. Odhelius laconically concluded, "Time will tell what will come out of it."[62]

It was part of the chymical elite of the German-speaking world that Odhelius discussed with such distance. Odhelius's growing knowledge permitted him not only to hold his own in conversations with other chymists but also to discuss other chymists and form his own opinion about them. His attitude, however, was not one of skepticism toward the chymical enterprise as a whole. Rather, it was the result of hard-won hands-on knowledge gained through his close encounters with wandering chymists and their often extravagant claims. Odhelius was learning how to separate the gold from the dross in the social world of chymistry. He also gained experience in how to present and conduct himself and choose topics of conversation. This too was valuable knowledge that would come in handy on his further journeys to other parts of Europe. His meetings with experienced and famous chymists also gave Odhelius

access to better-quality gossip. For example, the well-traveled Starcke
had informed Odhelius of the widely circulating rumor that the French
king had imprisoned a chymist who was employed in the manufacturing
of gold for the French crown.[63]

His opinions were inspired by his reading of Johann Joachim Becher's
recently published book *Närrische Weissheit*, in which some of these men
were mentioned and criticized. In a letter Odhelius referred to Becher's criticism of Marechalli, who was mentioned as one of several "publicly known
con-men and sophists." Odhelius also moderated Becher's low opinion of
Kraft by providing Hiärne with the information—probably from Kraft
himself—that Becher had "most of his own inventions from him."[64]

Odhelius and Hiärne were indeed deeply critical of many of their
contemporary chymists. Their criticism, however, was articulated on
moral grounds and should be distinguished from their theoretical positions on chymistry and on the viability of transmutation. As we have
seen, Hiärne was a Paracelsian, and at least until about 1690 Odhelius
was as well.[65] Hiärne's deep reverence for Paracelsus is evident in his
1709 pamphlet *Defensionis Paracelsicae Prodromus*.[66] In Hiärne's view,
Paracelsus was a true adept. He had known how to make gold, but he
was first and foremost a philosopher who had sought to open the innermost, hidden secrets of nature.[67] Paracelsus was also personally important to Hiärne. He claimed that Paracelsus was the source of almost
all of his knowledge of nature and the largest part of his knowledge of
medicine. Nevertheless, Hiärne's text also shows that he was familiar
with more contemporary texts. Among those whom Hiärne claimed had
understood Paracelsus were sixteenth-century chymical notables Adam
von Bodenstein, Alexander von Suchten, and Michael Toxites, but also
more contemporary chymists such as George Starkey.[68] It is obvious that
Hiärne was well read in the central tradition of transmutative chymistry.
He also held a clear opinion on who were the major opponents of this
tradition.[69]

In 1684 Odhelius penetrated the symbolism of chymical language.
"To give an appearance [of knowledge] and to describe the matter of
the stone (*materia lapidis*) in theory is easy enough," he said. What body
could one not dissolve, and then "apply the brother, sister, and the cup
of love in the mountains of the chymists, the babe in the womb, the
dragon, his tail and strong water, black blacker than black et cetera."[70]
The passage echoes the work of Michael Maier, but by itself it is little
more than a compilation of chymical tropes that give little indication of
the underlying processes that Odhelius may have witnessed or participated in performing. Many chymists worked with solutions of metals

in acids and other solvents, including vegetable ones, such as distilled wine, vinegar, and tartar. As Lindroth has shown, Hiärne too was preoccupied with such experiments. But it cannot be said with any certainty that Odhelius was referring to this kind of work.

Odhelius's newfound knowledge permitted him to draw his own conclusions about the search for the stone. No one could reclaim the great secret without God's particular grace. Therefore there was no need for secrecy concerning the composition of the stone. Ultimately it was God who decided who would come by it, and hence it was he who stood as a guarantee that it would not come into the wrong hands.[71] Odhelius's further conclusion is all the more surprising, as he took this as a reason to stop searching for the philosophers' stone altogether: he rejected chrysopoeia, arguing that it was a form of hubris. Since God's particular grace hardly was something that one could expect to receive, it would be better to be content with the discovery of some especially effective medicines. Even in this area there were many problems, and which medicines should be chosen? Odhelius doubted that there was a universal medicine that could heal and preserve both body and mind. There were plenty of such medicines in books and in heads—that is, in theory—but he had as yet seen none used in practice. One of the few things that seemed to be certain in this matter was that the human spirit must be of the same nature as the sun and as gold. For this reason, there were some extraordinarily effective gold-based medicines, such as fulminant gold (*aurum fulminans*) and sun tincture (*tinctura solis*). Another medicine that was possibly such a panacea was sulfur anodynum, although it only seemed to be effective against certain types of disease.[72]

A year later, in 1685, Odhelius extended his skepticism about the possibility of finding a panacea to that other central practice of chymistry, gold making. He also positioned himself theoretically against the rationalist physics of his day, including Cartesian atomism. It was useful, he stated, to have a hypothesis about the generation and improvement of metal. But prior judgments in these matters were of no value. One should always inspect each process in practice. It was a complete waste of time to do as, for example, Descartes in his *Creatio Mundi* and Becher in his *Separatio Chaotis*, that is, to begin with a discussion of the primary elements and arrive from that direction to the matter of chemical composition. Rationalist speculation about the creation of the material world was of little or no help for those whose aim was to understand specific chemical processes.[73] Chymists, on the other hand, had the right methods for going about the problem, but this did not necessarily mean that it was possible to achieve their ultimate goal to multiply metals. Chymists worked with solutions of metals and sought to find out how to obtain the prime matter that creates all things (*primam materiam seminalem*) through their laboratory arts. The reason was that nature composed perfect metals through the combination of various principles, while art sought to outdo her, by multiplying the seminal force that creates gold infinitely. Nevertheless, the best book of this type was Espagnet's *Enchiridio Physicae Restitutae*.[74] Although praised by Odhelius, Espagnet's book was nevertheless not primarily concerned with practically oriented, operative chymistry. Rather, it was concerned with chymical philosophy in the context of Neoplatonic cosmology.[75]

Odhelius's position was an advanced one that proceeded from intimate knowledge of contemporary matter theory and chymical experimental methods. He was no longer of the opinion that transmutation processes were simple occurrences that could be brought about through simple means. For example, he was skeptical of a common transmutation story in which transmutation was said to have occurred in nature: an iron ax turned to copper was to have been found in vitriolic water in a copper mine. It would be interesting to know, he wrote, whether the ax had turned into solid copper, or whether it had undergone only a superficial change.[76]

He would not veer away from the course he had taken. By 1690 he had dropped any hope of gold making altogether and stated that he was "confirmed in his thoughts" to "let go of all thought of chymically profitable works." It was much better, he stated, to stay with the profession that one had been assigned by God.[77] Hard work would not grant divine favors and was not enough to gain the philosophers' stone. Whether Odhelius intended it or not, his new position contained an implicitly authoritarian and recognizably Protestant critique of chymistry's disruptive social potential.[78] God was invoked, but it was one's king and one's betters who were to be obeyed. In this sense Odhelius's views were highly compatible with the authoritarianism of the absolute monarchy that he was primed to serve. Chymists, like everyone else, should submit to the worldly authorities and toil away at the incremental improvement of society as defined by their betters.

Although Hiärne never gave up his dreams of gold making, he was not the easily fooled believer portrayed by some authors. There were instances when Hiärne played the role of skeptic. When discussing an author named Faber in 1684, Odhelius said that Hiärne was too severe in his opinion. Faber had, after all, been honest in his manuscripts and had "shown the way to many things that cannot be found with others [in other texts] . . . but whether he was an adept, only God knows."[79] Neither was Hiärne particularly old-fashioned, which he also has been accused of.

Around 1700, the Paracelsian tradition was still thriving, but it was also increasingly on the defensive. Hiärne's work clearly shows this trend. During the period, Hiärne and his assistants investigated reactions and precipitations of a great number of substances. The substance to be investigated would be dissolved in a liquid. Then, a number of reagents would be added. Precipitations and other changes would be noted. According to Lindroth, who made a thorough study of Hiärne's laboratory journals for the period 1699–1707, the aim of these investigations was to show how the chemical principles, and in particular modifications of the principle of salt, made up all material objects. In part this work was intended to produce evidence that would counter the investigations of Boyle and Jan Baptista van Helmont, who questioned whether the Paracelsian *tria prima*, the principles of salt, sulfur and mercury, really were the underlying elements that built up all matter.[80] As a Paracelsian, Hiärne regarded the tria prima as central chymical concepts.[81] Hence it should come as no surprise that the publication of Boyle's *Sceptical Chymist* seriously disturbed him. As shown by Principe, Boyle's book should be understood as an attack not on transmutative chymistry but more specifically on the Paracelsian chymical tradition of which Hiärne was an exponent.[82] In his *Parasceve*, Hiärne wrote that he wanted to continue his work undisturbed but was unable to do so because of Boyle, the "Misochymicus," who "grasps my arm, asks me to wait and answer his objections and proposals through which he wants to overturn my entire work of precipitations, and show that the precipitations and the changes of colors are a thing that may please the eye but not the philosophical mind; meaning that all of it is a juggler's game and a jester's play." The great Helmont was, according to Hiärne, of a similar mind, aiming to make the chymical principles nonexistent.[83]

Hiärne was deeply engaged in the research of his contemporaries, so it is necessary to moderate Lindroth's and others' claims that Hiärne was old-fashioned. Although Hiärne stated that most of his knowledge came from notables such as Paracelsus, Pythagoras, and Plato, such claims must be treated with caution. We should understand him as an early modern humanist who tended to equate antiquity with value (let us not forget that Hiärne was also a poet, playwright, and linguistic scholar).[84] Hiärne's statements about his dependence on Paracelsus should be seen in light of his high regard for knowledge of the past. Many of the ultimate goals of medicine and chymistry (e.g., the philosophers' stone, the universal medicine and a number of lesser secrets) were heirlooms, passed on from generation to generation. There were also, however, a number of later authors (and contemporary chymists) who made valuable contributions to the corpus of chymical knowledge. Because chymistry was much more than just a textual tradition, it needed to be learnt as laboratory practice. A student had to do his own laboratory work, and engage with and learn from living knowledge-carriers, that is, other chymists who to greater or lesser degree embodied the practices that came with the knowledge. Hiärne undoubtedly was part of a contemporary tradition and no more old-fashioned a chymist than, say, Isaac Newton.

We have now discussed rather thoroughly Hiärne's and Odhelius' views of the central goals of chymistry. That is, the quest for the philosophers' stone and the possibility of finding a universal medicine and of creating gold. As his knowledge of the chymistry of the apothecaries and of the art of assaying grew, and as he gained a thorough familiarity with both chymists and chymistry, Odhelius expressed an increasing skepticism of realizing these goals. This skepticism influenced his decisions but did not mean that he would openly contradict or waiver in his support for Hiärne, his main patron and benefactor. Hiärne, on the other hand, seems to have disregarded the skepticism of Odhelius. He continued to place his trust in the younger man, and pursued the goals of transmutative chymistry and Paracelsian medicine for the duration of his life.

Chymistry in the Field

In the beginning of this chapter I stated that many seventeenth-century rulers assumed that chymists possessed useful knowledge that could yield economic benefits. They could produce all kinds of inventions and medicines, although they did not always value this knowledge. As Odhelius had remarked about the chymist Starcke: he had knowledge of almost every unusual art that could be learnt in Europe, but he valued nothing but gold making.[85] For the Swedish mining administration, the opposite was true. Gold making and the search for universal medicines held limited appeal, whereas the improvement of manufacture of goods was one of the central tasks assigned to the organization. The Bureau's traditional method of improving its knowledge base had been to hire foreign artisans.

Odhelius's thorough studies and Hiärne's unwavering patronage served Odhelius well on his return to Sweden. In 1687 he became an

FIGURE 5. Engraving illustrating a visit by Charles XI, Fabian Wrede, and other notables to the Sala silver mine in the summer of 1687. The image was probably composed by Arvid Karlsteen to be used on a silver medal that was struck the following year to commemorate the event. From an Uppsala dissertation defended by Lars Benzelius under the presidency of Petrus Elvius, *De re metallica sveo-gothorum schediasma* (Uppsala: Werner, 1703), verso, title page. Reproduction by Gertie Johansson, Hagströmer Library, Karolinska Institutet.

assessor extraordinary of the governing board of the Bureau of Mines. The following year he was promoted to the more prestigious position of head of a mining district, and in 1695 he was promoted to a regular assessor of the governing board. The following year, when Hiärne was appointed to the time-consuming task of president of the Collegium Medicum, Odhelius took over many of his duties at the Bureau. Hiärne, however, continued as head of the chymical laboratory.

Now that the Bureau had access to its own trustworthy chymists in Hiärne and Odhelius, various schemes were set in motion to use their knowledge for the improvement of the mining business. After all, Hiärne and Odhelius should be able to distinguish the bad from good in their own profession.[86] Most of these experiments were initiated during the presidency of Fabian Wrede (1685–94), who seems to have been a steadfast supporter of Hiärne's endeavors, although he, according to Hiärne, needed some persuasion as he "did not understand [chymistry], especially in the beginning."[87] Wrede's presidency and the following years up to the death of Charles XI in 1697 can be described as the heyday of Hiärne's influence at the Bureau. In addition, he was the favorite physician of the royal family and enjoyed its direct patronage, which resulted in his appointment as president of the Collegium Medicum.

Chymical reform efforts focused on the great copper mine of Falun. Although the mine was a legendary source of wealth during the seventeenth century, its yields were slowly diminishing. Effects were clearly visible in the decades around 1700, but the problems would increase as the eighteenth century progressed.[88] Because of the diminishing yield, the miners of Falun had taken to roasting the ore twice. In the first process, called pit roasting, the raw ore was crushed into smaller pieces and heated in an open-air pit for two to seven weeks. The roasted ore was then smelted in a furnace, and the smelted product was crushed and pit roasted a second time. Finally, the smelting chamber of the furnace was rebuilt at a smaller scale, and the material was smelted again. The entire process usually took two to three months, although one, two, or all of the first three steps could be omitted if the ore was very rich in copper. The resulting coarse copper—usually about 90 percent copper—was bought at a set price by the Crown and sent for refining to the state-owned Avesta works, where, smelted a fourth time, it was refined to a copper content of 97–99 percent, with traces of gold and silver.[89] As this process was lengthy, laborious, and expensive, the Bureau looked abroad for alternative methods. Eventually, however, it had to acknowledge that the foreign methods had been developed to be used on richer ores and were difficult to adapt to the local conditions at the mine of Falun.

From the point of view of the transmutative chymists who made their living around the mining business, the smelting process was an act of transmutation in which the ore was changed from a coarse and mixed state into pure metal. The idea that it was possible to chymically augment the process and receive greater yields was not far-fetched. Indeed, Paracelsus himself had argued that the secret of how iron was transmuted into copper had already been made manifest to humankind by God.[90] Urged on by Hiärne, the Bureau invited a number of foreign smelters and chymists to improve the copper smelting at the Falun mine.[91]

First out, in 1685, was the Italian Francisco Maria Levanto. Levanto's method does not appear to have proceeded from chymical theories. It was an application of methods that he had learnt at other European smelting works. In the end, however, it was deemed unsuitable. The ore of Falun was poor, with a very low copper content, and Levanto's method was only suitable when the ore was richer.[92] Levanto was followed in 1688 by the German chymist Johann Wendelin Schiumeier. Schiumeier conducted his trials under the watchful eye of Odhelius, at the time master of the mine in the district of Nya Kopparberg. Hiärne, at the height of his power, wholeheartedly supported Schiumeier, and in 1689 the man was appointed assessor of the Bureau. But the implementation of Schiumeier's process encountered difficulties. It was also becoming apparent that Schiumeier was an agent of someone else, the colonel Adam Friedrich von Pfuhl, of Helfta, near Eisleben. Pfuhl was the inventor of the process and had set up a consortium to send Schiumeier to Sweden.[93] In 1689 Hiärne himself traveled to Helfta to talk with Pfuhl and to learn his process. But Pfuhl was unwilling to part with the secret. Simultaneously, Hiärne's absence from court made his favored position vulnerable. His influence and reputation became an issue when his association with the foreign project-makers was used to slander him, and he had to return home quickly and without the secrets he wanted, in order to protect his interests.[94]

The Bureau instructed Odhelius to go to Helfta, gain Pfuhl's trust, learn and evaluate his processes, and report all his findings to his superiors on his return to Sweden.[95] He arrived in May 1690 and would spend the summer together with Pfuhl. Odhelius could substantiate that the colonel knew how to conduct fine operations on both medicine and metals, and they were soon engaged in adapting his operations to the larger scale. They developed two methods. The first was a new method to treat the ore, and the second was a new type of smelting furnace. In total, the new processes gave a slightly increased yield of copper, compared to other methods. Odhelius reported to the Bureau that Pfuhl's methods worked as he had promised.[96] He also seems to have gone along with Pfuhl's interpretation that the process was a successful augmentation. Despite the adaption of the chymical operation to the industrial scale, however, Odhelius did not think there was any economic potential in the new methods. The production costs were too high in proportion to the benefit. He did not recommend the methods for use in Sweden.[97] "This work will be forgotten by itself," Odhelius claimed, and "will serve as a warning from the Lord, that we should not try his patience and expect him to provide us with extraordinary means of sustenance."[98] Soon there was a consensus at the Bureau that Schiumeier's trials too had failed. Renewed attempts to invite foreign smelters and chymists were made during the 1690s, but the Bureau's support for these projects slowly eroded.[99] The trials came to a stop in 1696, when Johann Kunckel von Löwenstern left the country.[100]

In this undramatic way, Odhelius's travels with chymists came to an end. When he first set out in 1683, he had enthusiastically reported that he would investigate rumors related to him by a French chef in Ultenaw, that there were honest and god-fearing adepts in Hamburg.[101] In 1692 he traveled through Hamburg. Noting that Pfuhl was there, he quipped that the colonel most likely traveled in some *Chymisterie*. Odhelius declined to meet him.[102] By that time, he was no longer interested. What was more, he had other things on his mind. He had learnt about the recently developed practices of experimental philosophy and had his mind's eye set on England.

In Defense of Transmutation: Urban Hiärne's Final Years

On April 5, 1697, Charles XI died at Stockholm Castle. He was succeeded by his young son, Charles XII. With the new monarch came new
policies. It was by no means certain that Hiärne would retain his privileged position at court and in the administration. Hiärne had successfully
integrated chymistry in the institutional environment at the Bureau, but
toward the final years of the seventeenth century, the prospects for transmutative chymistry's continued existence there looked increasingly bleak.
As we have seen, even Odhelius had turned his back on the great work.

A further complication was the election of Didrik Wrangel as president of the Bureau in 1694. Hiärne claimed that Wrangel hated chymistry and curious studies, and persecuted them in secret. Maybe this was
true, or maybe the issue was more personal. In any case, Wrangel did
not appreciate Hiärne at all.[103] Charles XII's assumption of the throne
presented Wrangel with an opportunity to remove Hiärne from the Bureau. On his assuming of power, the young king was presented with
detailed information on the state of the kingdom. The Bureau's, and
Wrangel's, contribution was a thoroughly written memorandum on the
noteworthy developments that had taken place in Swedish mining during the reign of the late king. The report was sent in by Wrangel, who
carried ultimate responsibility for it as president of the Bureau. He had
co-authored it with four mine councilors and assessors, among them
Erich Odhelius. That Wrangel was directly involved in its writing is evident from the text. While the other authors are occasionally mentioned
in the third person, Wrangel at one point stepped out of the authorial
anonymity to speak in the first person.[104]

Hiärne was presented at a clear disadvantage. He was described as
someone who did more or less as he pleased. In a description of the chymical laboratory, the report gave an extensive account of its associated
costs, and of the many measures taken by the Bureau to support it. The
report went to great pains to show that the laboratory never really had
become a part of the Bureau but was an independent unit of doubtful
value. The Bureau had little real knowledge of what Hiärne had done
in his laboratory, as it had only received information from him when
he was in need of more money. The little knowledge it had came from a
report by Hiärne stating that a number of experiments on plants, trees,
and minerals had been conducted. The Bureau surmised that from these
experiments "something useful could follow, in the long run."[105]

The harsh judgment was extended to Hiärne's many other activities
and his longtime association with the Bureau as an assessor. Hiärne was

singled out as driven by emotion rather than a sense of duty. In 1685 he had made a journey to investigate a report about a rich find of copper ore in the province of Härjedalen. The Bureau's account stated that Hiärne had been given the assignment since he "fancied taking a trip" to investigate the minerals of the area. Although a copperworks had later been established at the site by a group of investors, the report carefully emphasized that the works were unlikely to yield any profit.[106] Praise was withheld even when no fault could be found with Hiärne's activities. Particularly striking is the report's narrative of Hiärne's initiative involving a general survey of the natural resources of the Realm in 1693. The initiative fitted well with the general policies of the Bureau and had few if any chymical connotations. It was nevertheless reported with a neutral and cautious tone.[107]

Chymistry had held a place not only in Hiärne's laboratory but also in several attempts by the Bureau to improve smelting processes. It was in this area that Hiärne's knowledge intersected most strongly with the needs and interests of the Bureau of Mines. In this area too, Hiärne's efforts, and those of the chymical projectors, were described in an unflattering light.[108] The tone of the document was echoed also in the protocols of the governing board of the Bureau for the year 1697. On several occasions, Hiärne applied for, but was not granted, funds to repair the laboratory and to buy materials for the making of medicines. On one occasion this request generated a discussion. In a meeting of the governing board Odhelius tried to stand up for his patron and read aloud to the president the royal edict concerning the funding of the laboratory. When hearing this, the president claimed he had never heard or seen this edict before. He continued by asking for a written report from Hiärne in which he should delineate how he had used the funds he had already received and show of what utility the laboratory had been so far. Furthermore, it was suggested that Hiärne should begin to hand in reports about the laboratory four times a year.[109]

His harsh judgment of Hiärne and chymistry did not mean that the Bureau's president damned natural philosophy in general. The account to the king contained a discussion of the role of knowledge in the organization. It made perfectly clear that the knowledge of how to manage mining works should be understood in the context of natural philosophy rather than that of crafts. The extraction of minerals had come to be regarded as a "noble science" pursued by "capable persons." It is equally clear that this knowledge was considered in a primarily utilitarian context. The mining works were important because they were the source of much of the Crown's wealth. Capable and well-educated men were needed *because* they could ensure the continued prosperity of the mining works.[110] For these reasons the Bureau had taken three measures to encourage young people to study the mining sciences. First, it granted traveling scholarships to selected promising persons.[111] Second, it accepted auscultators, young men who were permitted to be present at the Bureau's main office, and to travel the country in order to learn about the mining works. Third, it was in the process of setting up a mineral cabinet of Swedish mineral specimens. The mineral cabinet was to be an aid for young persons in pursuit of knowledge about minerals, but it was also to be used as a support for discussions at meetings.[112] The mineral cabinet, however, seems to have come to nothing. Both Hiärne and his successor to the laboratory, Magnus Bromelius (ennobled von Bromell in 1726), possessed private mineral cabinets, but the Bureau did not have one of its own until, at the earliest, the late 1720s.[113]

The final decades of the seventeenth century had seen a remarkable change of policy at the Bureau. Clearly, the concept of mining sciences was well established in 1697, and well-educated officials were perceived as vital. Mining knowledge was no longer to be associated with crafts but with natural philosophy. Officials were no longer to be brought from outside but were to be educated internally. Lacking a detailed analysis of Bureau protocols during this period, we should not credit Hiärne with too much influence over this change. But it is obvious that the measures taken up by the Bureau were in agreement with his ideas, both on the importance of applying natural philosophy to mining, and on the proper education of the mining official, as executed in the education of Odhelius. Hiärne's ideas had been both institutionalized and disassociated from him.

Although it was not spelled out in the report, transmutative chymistry was out of favor. It had not been discredited on theoretical grounds but on practical ones. Recall Odhelius's final judgment of the copper process of Colonel von Pfuhl. The colonel's transmutation of copper had worked but was not deemed to be economically feasible in Falun. The disassociation of Hiärne's laboratory from the Bureau's day-to-day activities was not really surprising. The type of research that was done in the laboratory no longer carried any particular promise of monetary returns. Furthermore, Hiärne would soon lose his main ally and friend at the Bureau. Odhelius died in 1704, at the age of forty-three. Without Odhelius, there was no longer a mediator between Hiärne and the rest of the governing board of the Bureau, which now consisted largely of men infused with a degree of skepticism toward transmutative chymistry. In a 1706 overview of what he had achieved in his laboratory,

Hiärne even distanced the sensible chymists (and himself too) from gold making and emphasized that his old patron, Charles XI, had supported chymistry for other, more praiseworthy reasons.[114] The usefulness of the laboratory was not seen in the context of transmutative chymistry but in the context of inventions such as the compass and gunpowder.[115] Despite his lip service to the growing anti-transmutative sentiments, Hiärne continued his quest for the great Arcanum of gold making.[116] During the final decades of his life, he fought a number of rearguard actions in defense of his art.

Another problem for Hiärne was that the kingdom's most important patron, Charles XII, did not show much of an interest in transmutative chymistry. He favored mechanical project-making, and in particular such inventions and innovations that would aid in the war effort of the long and grueling Great Nordic War (1700–1721). The monarch's views became an issue in the von Paykull affair, which played out in 1707. Otto Arnold von Paykull was a military officer in the service of Saxony. Charles XII ordered him executed in 1707, after he had been taken prisoner during a battle between a Saxon and Swedish army. The reason was that he had been born in a Baltic province under Swedish rule and therefore, when he turned his arms on Sweden, he became a traitor according to the country's law. Paykull, however, was also a chymist of some renown. He claimed that he could produce large quantities of gold for the Swedish crown and also teach his art to a Swede, in exchange for a pardon. Hiärne involved himself. He was convinced that Paykull spoke the truth about his abilities. Slipping out of his hands were the opportunities to learn the great Arcanum and to redeem his art in the eyes of the king by producing gold for the war effort. After Paykull's execution, Hiärne wrote a short treatise, a testimony that "Paykull truly was an adept, or that he knew how to make gold"; this treatise was not printed but appears to have circulated widely in several copies.[117]

Hiärne had now, with a certain amount of recklessness, criticized the king's judgment.[118] This was a dangerous business. Furthermore, the king had made threatening remarks about chymical adepts in general during the affair, and new difficulties were also brewing at the Bureau. Hiärne had learnt that there were plans in some quarters to propose a certain Schering Rosenhane for president of the Bureau. He had discussed physics with Rosenhane at times and regarded him as "devilishly" arrogant. Hiärne sought support from his patrons and went so far as to sound out the possibility that the laboratory be removed from the Bureau and made part of the king's personal household. The final outcome was that Rosenhane was not elected, and Hiärne remained.[119]

The following year of 1708 left no respite, as the young provincial doctor Magnus Gabriel von Block published a pamphlet containing, among other things, sharp criticism of the character and work of Paracelsus.[120] Block seized on one of Hiärne's arguments from the Paykull affair, that Paykull's offer was the coming into being of a well-known prophecy of the return of the Lion of the North, which would herald the triumphal victory of Protestantism over its enemies. According to Lindroth, Block did not really attack Paracelsianism but perceived Paracelsus's teachings as composed of older and more reliable sources. It was the person Paracelsus that he sought to discredit. Although published in the provincial town of Linköping, Block's pamphlet was to be mentioned in the *Journal des Scavans* in 1709 and was translated into German in 1711. Hiärne's response was to amass his still considerable influence and threaten Block into submission.[121]

Hiärne was growing old, surrounded by opponents and younger critical talents, and losing friends. Odhelius had died in 1704. In a letter of 1718, the bishop Jesper Svedberg commented that Hiärne formerly had been his good friend and brother but that he lately had heaped abuse upon reputable men "due to his arrogance, in claiming to know everything about everything, which has always been, and still is, one of his peculiarities."[122]

The heir to Hiärne's positions was to become the physician Magnus von Bromell. He was appointed assistant director of the laboratory in 1719, and the following year he was made an assessor at the Bureau. Bromell also took over Hiärne's positions at court and at the Collegium Medicum. As Bromell was thirty-eight years younger than Hiärne, there were obvious differences in the outlook of the two men. The son of the town physician of Gothenburg, Bromell had studied medicine for three years at the University of Leiden with some of the most prominent teachers that Europe could offer.[123] After a short stint in Oxford in 1700, he returned to Leiden in time to take classes with the newly appointed Herman Boerhaave and to defend two theses. I have not been able to locate them, but one of them bears the tantalizing title *De Non Existentia Spirituum* (Concerning the nonexistence of spirits). It testifies, if nothing else, that Bromell was no stranger to the debates concerning the relationship between spirit and matter that took place during this time.[124] After completing his studies and conducting a European journey of seven years, Bromell moved to Stockholm to start a medical practice. Just as Odhelius before him had gathered collections associated with his trade, Bromell amassed anatomical preparations, books, and naturalia during his travels.[125] Part of his collection was presented in the

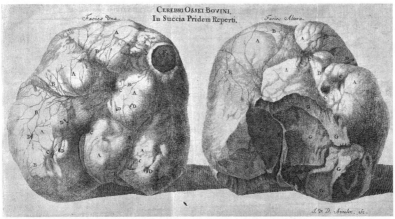

FIGURE 6. Ossified brain of a cow. From Magnus von Bromell, *Lithographiae Svecanae: Specimen primum, Calculos humanos, variaqve animalium concreta lapidea exhibens, Juxta seriem atque ordinem, quo in Musaeo Metallico Bromeliano servantur* (Uppsala, 1726), foldout between pages 22 and 23. Reproduction by Esbjörn Eriksson, National Library of Sweden.

two-volume *Lithographiae Svecanae*, a work that shows clear influences from the Leiden school of natural history with which Bromell was so familiar. It also transplanted Dutch collecting, with its focus on anatomy and exotic plants and animals, to the realm of mineralogy. The books dealt with petrified specimens from the human body, and the animal and vegetable realms.

The *Lithographiae Svecanae* pointed to the continuity between Bromell's mineralogical interest and medical practice, while also functioning as a handsome presentation of the contents of his personal museum.[126] At the time of his death (1731), Bromell possessed one of the most significant collections of naturalia in Sweden.[127]

Undoubtedly, Bromell used his collections to establish himself as one of the most important *curiosi* of Stockholm. Not only was he to take over most of Hiärne's prestigious positions, as well as some additional ones, but he also continued his medical practice.[128] Although Bromell seems to have been the right man for the job of reestablishing the Laboratorium Chymicum, he obviously had overreached in his ambition. The laboratory was left to decay, and eventually the directorship was handed over to the guardian of the mint, a goldsmith by the name of Mikael Pohl. In practice, it was run more or less independently by Hiärne's aging laboratory assistant, Johan Sahlbom, until his death in 1723.[129] Although Bromell did not put much of a mark on the Bureau, his education and interests show clearly where things were heading. Natural history as well as the new experimental philosophy would set the agenda at the Bureau for the century to come, not Paracelsian natural philosophy or transmutative chymistry.

: : :

The Bureau of Mines had been established as a means for the state to control and improve the lucrative Swedish mining business. Its first officials had been immigrant, mostly German, artisans. Although the early Bureau had little interest in natural philosophy, this would all change as Hiärne was attached to the Bureau. Hiärne and his disciple Odhelius would introduce a number of key innovations.

In the context of the Bureau of Mines, Hiärne should above all be seen as an institution builder. The institutionalization of Hiärne's chymistry as a part of the Swedish mining administration established a contact zone in which the knowledge of Swedish and German artisans and chymists was appropriated and transformed. It became a form of knowledge suitable for high-status mining officials. This contact zone continued to exist and expand long after Hiärne's demise, and long after the general abandonment of Paracelsianism at the Bureau. In addition, the introduction of the institution of auscultators enabled young men to spend time at the Bureau, study the mining sciences under the tutelage of experienced officials, and listen in to proceedings. Also of great importance was their access to the Bureau's archives, where they were permitted to copy important documents.[130] They could, for example, compile excerpts of the travel journals of their predecessors at the Bureau. In the 1730s, young mining officials could conduct virtual foreign journeys, accessing, for example, the journeys in the Holy Roman Empire of Erich Odhelius (1692), Jacob Hertzen (1697), Samuel Buchenfeldt (1697), Thomas Cletscher (1701), Georg Brandt (1724–27), and Gustav Gabriel Bonde (1731) and compile them into a single volume for easy reference.[131] As the fame of the Bureau grew, its expanding archives would come to be considered an important resource by officials of the mining administration of other countries. In 1780 the Swedish-born Prussian mining official Johan Jacob Ferber asked his Swedish correspondent P. W. Wargentin to act as an intermediary and procure permission for him to have excerpts of documents made by the Bureau's officials. Wargentin had already supplied Ferber with printed Swedish chemical and mineralogical works, and willingly obliged this request also.[132]

Later generations of chemists would make themselves at home in the institutional structures created by Hiärne and Odhelius. They would access chymical writings in the archives of the Bureau. They would use Odhelius's journeys as a template for their own grand tours of Europe.[133] They would receive positions and salaries created for and by chymists. Nevertheless, for them, chymical theories and the practice of gold making would be regarded as ridiculous activities. Alchemy itself would for them become part of a backdrop against which their own version of Enlightened science could create a distinctive image for itself.

We have seen how the Holy Roman Empire, and in particular Saxony, was a preferred place to search for adepts and also to find skilled laboratory assistants. The Bureau of Mines sought to establish contact with knowledge holders by sending people to the places they frequented. Hiärne did not only send Odhelius but he himself traveled to Eisleben and other places. Another, and even broader, contact zone that the Bureau had access to was the Swedish state's well-developed habit of relying on foreigners. It ensured Hiärne a steady supply of assistants, and even though foreigners rarely were given salaried positions at the Bureau from the mid-seventeenth century onward, the expansive Swedish Realm saw a continuous influx of foreign smelters, chymists, and projectors throughout the seventeenth century and beyond. Access to the contact zones engendered by frequent travel and immigration of labor gave actors at the Bureau of Mines a distinctive advantage as producers of natural philosophical knowledge. These advantages were not shared by stationary actors, located at alleged centers of knowledge production. We can compare Hiärne with the London-based Robert Boyle. Boyle, too, sought chymical knowledge in the Holy Roman Empire but had to be content with interacting through correspondence and with such men as traveled to visit him in England.[134] This is not to deny that Boyle had advantages of other kinds. Toward the final decades of the seventeenth century, it became apparent to the Bureau's officials, as well as to many other Europeans, that exciting environments where one could learn new and curious things were to be found not only in the Holy Roman Empire, but also in the Netherlands and in England. It is to those places that we now turn our attention.

4

From Curious to Ingenious Knowledge

Around 1700, European natural philosophy was in a phase of transition. The curious, open-minded, and what has been called *baroque* knowledge tradition of the seventeenth century began to give way to other pursuits and interests. Experimental philosophy, mechanical project-making, and natural history collecting continued their advance as promising fields of inquiry. These emerging pursuits were united by the intellectual framework provided by mechanical philosophy.[1] Its central metaphor, "God is an artisan," encouraged a perception of nature as an extension of the sphere of human artifice. It was also preoccupied with power: nature could be explained as a collection of ingenious devices and clever inventions, each of which was intended for specific uses. This was a form of knowledge articulated from a utilitarian outlook and guided by a desire to improve, change, and take control.[2] It was an ingenious rather than a curious philosophy of nature.[3]

This chapter discusses the impact of these new intellectual currents. During the period ca. 1690–1730, mechanical philosophy became integrated into the activities of the Bureau. Mechanics, mathematics, and surveying came to dominate discourse. Soon, mechanical conceptions of the world would penetrate also into the sphere of chymical knowledge and transform the Bureau into a stronghold for a new type of mechanical chemistry.

It was a sweeping change, which came about through the Bureau's close encounters with other knowledge milieus: in particular, London, Uppsala, and Leiden. By taking these sites as their primary points of reference in the world of knowledge, the officials of the Bureau swiftly changed their intellectual orientation. Through traveling, studying, reading, and gathering of naturalia, the officials reshaped their knowledge and worldview, but also their perceptions of themselves. The chapter begins by revisiting Odhelius's travels in order to see how his views were transformed by his encounter with London and with experimental philosophy. I then show how mechanical project-making came to replace its chymical counterpart as the Bureau's primary tool in its attempts to improve the mining business. There was also a renewed and stronger interaction between the Bureau and Uppsala University, where there was a renewal of ambitions to establish the university as a center for knowledge in its own right. Nevertheless, chymistry remained somewhat separate from the circles of mechanical projectors and philosophers until the arrival of Georg Brandt at the Bureau. Brandt would introduce the new chemistry of Hermann Boerhaave at the Bureau and abandon the Paracelsian tradition altogether.

Erich Odhelius in London

It is now necessary to break the chronological narrative and backtrack to 1692, when Erich Odhelius was alive and well in Hamburg and thoroughly fed up with transmutative chymistry. His attitude had been shaped through personal encounters with several of the most renowned practitioners of the German cultural sphere. But underlying Odhelius's rejection of transmutation was also his encounter with new methods to pursue knowledge. He had by then spent a long time journeying in England, France, and the Netherlands. Odhelius was not unique in turning his back on Paracelsus and chymistry, and in a sense also on the Holy Roman Empire (and France), at this time. He is of interest, however, because we can follow in detail his change of views, and influences, during the period 1686–92.

In the beginning of November 1686 Odhelius arrived for the first time in London, on the lookout for everything that might be "curious."[4] He did not like what he saw. Rather than impressed, he was disturbed by the enormous financial resources that the English had at their disposal. This sense of irritation also colored his description of his obligatory visits to the great Robert Boyle. Odhelius's letters give the impression that he perceived Boyle as someone who lived off the ingenuity and resources of others, and could no longer produce valuable knowledge himself. As soon as Boyle had an idea in his head, Odhelius wrote,

> there were Mathematical machines, instruments, assistants &c, soon there at hand. He has much to which he should give thanks to strangers, because he is wise in his fishing. . . . Lately, I have not wanted to go to him, since he wanted so carefully to ask about this and that, which had cost me a lot of work [to learn]. He is old now, and would do well to die with a [good] reputation, and to cease to write further, because his things are not ingenii auctoris.[5]

Interestingly, Odhelius's critical observation regarding Boyle's use of assistants to execute his ideas has similarities to those made by Shapin in a 1989 paper. Hence, the impropriety of Boyle's invisibilization of his technicians was not lost on at least one of his contemporary visitors, who furthermore connected it to Boyle's preying on the knowledge of foreign visitors.[6] It was not just Boyle that Odhelius disliked, however. His impression of the English was squarely unfavorable, and when he left England it was with a sigh of relief.[7] He would not come back until 1691. On his second visit he profoundly changed his attitude toward England and the English. In his report to the Bureau, he praised England as the most wonderful home for "natural speculation and chymistry." Indeed, England had come to approach the level of Germany in the chymical sciences, as well as in the preparation of colors, medicines, useful household goods, and mechanical works. The king, the Parliament, and private benefactors all supported the English to advance in these areas.[8] In his letters to Hiärne the tone was even more enthusiastic. He told Hiärne that he admired the limited monarchy of the English. He was also deeply impressed by the beauty and bounty of the English countryside. It is evident that the core group of the Royal Society received him with great courtesy. He was presented at the society by a young Hans Sloane, spent much time with Robert Hooke, and claimed acquaintance with most of the Society's London-based fellows, such as Edward Brown, Martin Lister, Robert Plot, and Richard Beaumont, "all of who have treated me politely."[9] The reception of Odhelius in London shows that he had now mastered the gentlemanly social codes that were considered proper in the Royal Society. Indeed, he had internalized them to such an extent, that he on one occasion rebuked Robert Hooke for his unwillingness to communicate information.[10] Odhelius's formal report too contained a description of the Royal Society, which, it seems,

Odhelius saw as devoted primarily to the gathering of knowledge and to utilitarian improvement. It was even, he claimed, partly devoted to the gathering of information about mining enterprises within and without Europe.[11] It is difficult to say whether he described what he had been told, what he wanted the Society to be, or what he believed his superiors should hear.[12] In any case his description of the Royal Society resonated well with his own utilitarian interests and served to promote the type of foreign knowledge gathering that he himself was engaged in. In his letters to Hiärne it also comes across that he regarded the Royal Society as something of a patriotic achievement: "This society will contribute to England's eternal glory and [be of] no less use in the long run[.] [A]nyone who observes their method can perceive, that not only do they advance in speculative sciences, but [they] also truly contribute to the improvement of their country."[13]

The British were ascribed industrial inventiveness, utilitarianism, and patriotism, and were depicted as an example that the Swedes should emulate. The Royal Society made a particular impression: Odhelius even indicated that he had plans to set up a similar society in Sweden.[14] He was also beginning to internalize the experimental philosophy advocated by the Society and the demonstrations that accompanied it. His positive view of London shines clearly through in his report from Paris, of June 1692. To the extent that he found anything positive to report about Paris, it was that many in the city had adopted the New Philosophy: "For as long as I stayed in Paris, I visited one and another Curiosus, and I have seen what has been going on in their Academies[.] [T]hereby I have noted that the new Philosophy and experimental demonstrations have been accepted here, and are taught publicly to a much greater extent, than when I was here the last time."[15]

Concerning Parisian chymistry, he remarked that it was pursued with greater diligence than previously but that *charlatanerie* still ruled in Paris. The city's Collegium Medicum, the body that oversaw medical practitioners, was held in contempt, and its members were little more than charlatans themselves.[16] But the Paris chymists were not all bad. Odhelius noted that it was a good thing that Nicolas Lémery still gave his chemical courses. There was also one man who stood out: the recently appointed chymist to the Académie des Sciences, Wilhelm Homberg. Odhelius discussed the man with respect and said that Homberg had asked if he could recommend a German laboratory assistant. Apparently Homberg too had little to say for the local chymists.[17]

Odhelius also noted, with evident relish, that the French academicians had been expressly forbidden by the king "to conduct alchemical work"; in this way the king hoped to avoid rumors that his wealth came from gold making. The remark is interesting in that it clearly shows that Odhelius, by this time, had begun to identify alchemy with gold making, and distinguished it from general chymistry. Apparently Odhelius's turning away from transmutative chymistry corresponded with a renewed emphasis on experimental philosophy and the utility of the application of natural philosophy to industry and crafts. His report to the Bureau approvingly described how multiplication of gold had been made a felony in England, although this sadly had not prevented gold making from thriving anyway.[18] We may assume that his views on these matters, although formed at an earlier date, were reinforced by his encounter with the dynamic entrepreneurial milieu of London in the 1690s.

Enter the Mechanici

The chymists were not the only group of knowledge holders to stand out as especially notorious to early modern observers. Of similar fame and notoriety were the mechanical project-makers. Just like the chymists, mechanical project-makers traveled widely and were associated with the promise of huge financial rewards, or utter ruin. Alex Keller paints a vivid picture of the international scope and intensity of project making in seventeenth-century England: "German technicians and German investment drove the Company of Mines Royal. Dutchmen like Bradley and Vermuyden planned the draining of English marshes, and drew on Dutch investors for their support. French Huguenots were settled in the Fens; Italian engineers were everywhere, even in England, employed in fortification in time of war, on drainage or canal excavation in more peaceful times."[19] Similar to England, Sweden had been a technological backwater in the sixteenth century and became fertile grounds for foreign mechanical projectors in the seventeenth century.[20]

Much of the project making supported by the Bureau of Mines focused on improving smelting processes. Comparably less attention seems to have been devoted to mechanical projects.[21] A number of mechanics, *mechanici*, were attached to the Bureau. Like the other officials of the Bureau, they tended to be German-speaking immigrants—or their descendants—from the mining districts of central Europe, and they learnt their arts as apprentices of their seniors, who typically also were close relatives. This would all begin to change from about 1690 when Christopher Polhammar (ennobled Polhem, 1716), the most successful of the new breed of mechanical project-makers, entered the scene.

Polhem was the son of a German immigrant trader, builder, and artisan. A poor but obviously gifted boy, he became a student at Uppsala through the help of various employers and patrons. In 1690 he was engaged by the Bureau of Mines, and soon he would advance swiftly in the service of the state. Four years later Fabian Wrede ensured him a stipend that permitted him to embark on a two-year journey to European centers of invention, industry, and mining. In 1697 he was set up with a semi-independent Laboratorium Mechanicum modeled directly on Hiärne's Laboratorium Chymicium, and in 1711 he became a protégé of King Charles XII.[22]

Just as Hiärne's did, Polhem's position at the Bureau depended on his patrons.[23] Hiärne had stated that Wrangel hated chymistry and curious science, but there can be no doubt that he supported ingenious science. The 1697 report to the young King Charles XII sketched a striking, and likely intentional, contrast between Polhem and Hiärne. As Hiärne was denigrated, Polhem was spoken of with admiration and respect, and his mechanical innovations were described as highly useful. Furthermore, "quick in-geniuses" in general were to be encouraged to study at the Bureau.[24] When Polhem had been attached to the Bureau of Mines, the Bureau stated that "Polhammar is a native Swede, and has gained a good knowledge of both mathematics and mechanics proper, and is thereto gifted with an inventive and nimble *ingenio*."[25]

The word *ingenio* was also used in other descriptions of Polhem's character in official documents and letters of recommendation.[26] The shift of emphasis from curious to ingenious corresponded to a shift of emphasis from chymistry to mechanics in what by now had become the science policy of the Bureau (to use an anachronistic term). The report pointed clearly in which direction it wanted the young monarch to go. Transmutation chymistry, that high-risk, high-reward game, had little part in the report's vision of the future for mining.

The new policies of the Bureau had a concrete expression in the foundation of Polhem's Laboratorium Mechanicum. It was envisioned as a school for the teaching of students in mechanics, a laboratory for the conducting of practical and theoretical trials, as well as a permanent exhibition for models of mining machinery. Aside from Polhem himself, the laboratory would employ a carpenter, a blacksmith, and their two apprentices. It would also receive a substantial budget for the purchase of materials. In reality, however, Polhem was too occupied elsewhere, and much of the funding was withheld due to the outbreak of the Great Nordic War. The laboratory seems to have been reduced to a symbolic marker of status, connected to a sporadically paid state subsidiary for

Polhem's personal workshop. Again we can see the close parallel to Hiärne, whose laboratory at times was moved to his private quarters, and which was used to further Hiärne's own agenda of metallic transmutation. Because Polhem removed his laboratory to his rural property Stjärnsund and did not have a seat on the governing board, however, the Bureau tried to exhibit a greater degree of control over his enterprises than they did over Hiärne's. At times the Bureau would write sharp reminders that the laboratory was intended for the benefit of the public and not to be used to develop machinery for Polhem's private purposes. It is clear that Polhem neglected many of his duties as an official, and complaints against him mounted. His reputation would probably not have survived if his patrons had not dominated the governing board of the Bureau.[27]

In 1711, two timely letters of recommendation gained Polhem the personal attention of the king. Charles XII was at the time holed up in Turkey, where he had fled after the annihilation of his army at the hands of the Russians. In a series of letters from Turkey, the king ordered the domestic Swedish authorities to properly support Polhem in his work. The king was particularly interested in military inventions, and thus Polhem's ingenio was swiftly enlisted for the war effort. But the king's attention was a mixed blessing. Charles XII was despised in many quarters. When he was shot in Norway during a siege in 1718, persistent rumors had it that he had been dispatched by means of a silver button, fired from his own entourage by someone who held him to be magically hardened against ordinary lead bullets. After his death, swift plans were set in motion to facilitate peace negotiations as well as a political transition from absolute monarchy to parliamentary rule. There would be a massive shift in political practices. During the Age of Liberty (Frihetstiden, 1719–72), political power resided with Parliament, and the role of the monarch was reduced to an almost entirely symbolic one.[28] The new government did not appreciate the old favorites of Charles XII. Polhem was considered one of them, and for the time being, he fell out of favor.[29]

Embracing the Useful: Swedenborg in Uppsala and London

Medieval universities had primarily been church institutions. From the seventeenth century onward, an increasing number of (especially Protestant) European rulers began to take control of the universities in their realms and use them for patronage. That is, clients of the sovereign who had made their names at court were to an increasing degree awarded with university chairs. Professors took on a new role, that of clients of rulers or of influential politicians. The change coincided with a number of ideological changes at universities, as focus was redirected toward practical subjects conceived of as useful to the state.[30]

A further aspect had a direct bearing on governing: the universities were charged with educating state officials in useful knowledges and in cameralist economy. In the first half of the eighteenth century, Uppsala University was to become the main supplier of manpower to the central state apparatus that had emerged in Stockholm. Strong ties were also knit between individual Uppsala professors and Stockholm state officials and politicians.[31] A high road to public office was established. Many a career at the eighteenth-century Bureau of Mines began with studies at Uppsala and continued with a lowly and unpaid position as auscultator. A stipend from the Bureau, a rich family, or the position as supervisor of a young nobleman doing his grand tour would allow a journey abroad for a number of years. On his return the now not-so-young auscultator would usually be awarded a salaried position and subsequent promotion to higher office as time went by. Those who reached the position of assessor did so after serving on average twenty-three years at the Bureau.[32]

This new circulation of people between various places where knowledge was learnt, used, and produced depended on the existence of an infrastructure that facilitated education in mining matters. This infrastructure was partly formalized, as shown by the institution of accepting auscultators, permitting them to study at the laboratories of the Bureau, and granting them stipends. It was also partly informal, as recruitment relied on personal—and family—networks. The close links between the Bureau and Uppsala University relied both on family connections and on the maintenance of regular exchanges through which philosophical agreement and shared viewpoints were established.[33] Of importance is also the fact that a position at the Bureau had emerged as a good career choice. In the period 1715–25 there were about eighty salaried positions at the Bureau at any given time; thirty of these were attached to the Stockholm office and the remaining were based in the mining districts. In addition, there were about fifteen nonsalaried auscultators attached to the organization, and on average, three or four fresh auscultators were admitted each year.[34] Of the fifty assessors who were appointed to the Bureau's governing board during the years 1708–94, all but three had begun their career at the Bureau as auscultators.[35] Hence, individual auscultators could look forward to rather good career prospects, especially if they had good connections or relatives who were already working at the Bureau.

In 1710 Polhem began to correspond regularly with a group of professors and other learned men at Uppsala. The group initially called itself the College of the Curious (but it would soon drop the designation *curious*) and had gathered around the university librarian Eric Benzelius.[36] At its core were also the professors of astronomy and mathematics Pehr Elvius and Harald Wallerius.[37] The contacts were of mutual benefit. A letter of recommendation from the professors would help Polhem in his efforts to gain the attention of the king. Harald Wallerius's son Göran and Benzelius's nephew Emanuel Swedenborg (Svedberg before his ennoblement in 1719) became Polhem's students in mechanics at Stjärnsund. Both of them would later become assessors at the Bureau.[38]

The College of the Curious discussed primarily natural philosophy and mechanics. The participants in the group were Cartesian adherents of mechanical natural philosophy and of the geometrical method.[39] There can be no doubt that the circle was engaged in reinterpreting nature in the terms of mechanical philosophy. As good Cartesians, the members of the college adopted a highly reductionist approach to the study of curious phenomena. For example, the creation of metals in the earth, as well as strange sounds or lights, was interpreted as the consequence of mechanical and natural transformations.[40] The members of the College, however, were no slavish followers of Descartes. They tried to keep themselves updated on the latest theories, in particular in celestial mechanics and the new mathematics that had been developed by Isaac Newton, Gottfried Wilhelm Leibniz, Jacques Bernoulli, and others.[41] In 1710 Elvius and Polhem discussed Newton's *Principia*. Both would reject it. Elvius regarded it as "pure abstraction," and Polhem opted for another theory that did not presuppose the existence of gravity.[42] In a letter to Benzelius, he stated that he had conducted experiments that would change several of Descartes's opinions, while confirming others.[43]

The group's access to high-quality information would improve radically as Benzelius's nephew Swedenborg commenced his first foreign journey. The College of the Curious perceived the pursuit of knowledge as a patriotic project. Hence, collection and relaying of foreign information that could be useful for the state was a priority.[44] Swedenborg would perform a similar function for the College of the Curious as Odhelius had done for Hiärne and the Bureau a few decades before.[45] Arriving in 1710, he would immerse himself for two and a half years in the mechanical and entrepreneurial scene of London, before continuing to Paris and eventually back home. He dutifully relayed his impressions to friends and patrons. The young mechanicus had a contagious enthusiasm for England and for experimental philosophy:

This island, however, has also men of the greatest experience in this science. . . . I study Newton daily, and am very anxious to see and hear him. I have provided myself with a small stock of books for the study of mathematics . . . an astronomical tube, quadrants of several kinds, prisms, microscopes, artificial scales, and *camerae obscurae,* by William Hunt, and Thomas Everard, which I admire and which you too will admire. I hope that after settling my accounts, I may have sufficient money left to purchase an air pump.[46]

Somewhat later, Swedenborg, apparently on a shopping spree with too little money, asked his uncle if he would sponsor the purchase of an air pump. He also offered Benzelius to buy a full set of the *Philosophical Transactions.*[47] The pleas did not fall on deaf ears. On June 20, 1711, the Uppsala society noted in its minutes that the *Philosophical Transactions* had been purchased for the University Library. The society also wanted to have further information from Swedenborg concerning the price and description of an air pump, as well as additional information on English instruments and globes.[48] Gossip was also exchanged, as when Prof. Elfvius asked for the opinion of the Englishmen of Isaac Newton's *Principia.*[49] Paris, by contrast, held as little appeal to Swedenborg as it had for Odhelius two decades earlier. Swedenborg visited some old acquaintances of Benzelius, but did not report on seeing anything of interest to the College of the Curious.[50]

Swedenborg's journey, and in particular his encounter with London, had turned him into a mechanical projector, eager for glory and ready to serve his native country. He stated that he had conceived of fourteen mechanical inventions, among them a military submarine and an air gun, several pumps serving various purposes, a universal musical instrument, and "a flying carriage, or the possibility of remaining suspended in the air, and of being conveyed through it"[51] Wisely enough, the machines that Swedenborg eventually would choose to draw out in detail were pumps and machines to raise weight by means of water, as well as a new type of sluice, a steam engine, and two designs for rapid-firing air guns.[52] Pumps and sluices held more appeal to state officials than flying chariots, and the promise of improved guns that could turn the table in the struggling war effort would hold appeal for the warrior king Charles XII.[53] Indeed, Swedenborg's new English style of project making received a favorable ear with Charles XII. He was permitted an audience with the king through Polhem and soon became a personal client and protégé. The king also awarded him a position as assessor extraordinary at the Bureau of Mines.[54]

It seems, though, that it was a journal, the *Daedalus Hyperboreus*, that had gained Swedenborg the king's support. During Swedenborg's journeys, he had drawn up plans for a journal through which the ideas and inventions of the Uppsala society would be disseminated to a wider audience. The *Daedalus Hyperboreus* appeared in six issues during the years 1716–18. It was edited by Swedenborg, printed with the help of Benzelius, and featured Polhem's name on the title page. The essays it contained were primarily mathematical and physical, and written by Polhem and Swedenborg. Charles XII read it eagerly, and apparently kept the first issue on his table for three weeks, showing it to several visitors.[55]

Swedenborg's influence with the monarch gave him considerable leeway in his dealings with others, and now he took on old Hiärne himself, who by now had reached the venerable age of seventy-five. Swedenborg's father, Bishop Jesper Svedberg, was already engaged in a controversy with Hiärne over a suggested spelling reform of the Swedish language.[56] The son now came to the defense of his father, promising to send Hiärne a warning "to stop his impertinencies; because it is quite possible that someone may show up the puerilities and shortcomings in scientific matters, which he himself has had the daring to publish."[57] It was indeed a daring move, as Hiärne at the time was vice president of the Bureau and his superior.

The letter was sent in October 1718. On November 30, the king was shot. Swedenborg suddenly found himself without his most powerful patron. A new, and more tuned down attitude was in order. A year later, Swedenborg seems to have made several efforts to befriend Hiärne and asked his uncle Benzelius to "smooth the old gentleman down as much as you can."[58] The campaign seems to have succeeded, and soon Swedenborg was spending time in Hiärne's company. The older man would have approved of his desire to make himself "at home in mechanics, physics, and chemistry, and at all events, to lay a proper foundation for everything, when I hope no one will have any longer a desire to charge me with having entered the [Bureau] as one entirely unworthy."[59] The charge that Swedenborg had entered the Bureau as an unworthy favorite of Charles XII would haunt him for many decades. Many at the Bureau seem to have considered him an outsider.[60] He started to write prolifically in several areas. Some writing focused on mining technology and metallurgy, but he also produced a number of works on physics, writing in a late Cartesian tradition but bearing strong influences from Christian Wolff, and through him, Leibniz. In this way, Swedenborg secured a position as a respected mining official.[61] His weak position at the Bureau also forced him to seek new alliances. Now the lovers of mechanics, physics, and chemistry were contrasted with the assessors who specialized in mining legislation.

Swedenborg began to make himself at home. He befriended Anton von Swab, one of the Bureau's foremost chymists and assayers, and initiated a correspondence with the famous Freiberg chymist and mineralogist Johann Friedrich Henckel.[62] Swab had in fact studied with Henckel, who was one of Europe's most influential teachers of mining chymistry and mineralogy at this time.[63] During 1724 René Antoine Ferchault de Réaumur visited the Bureau, and a letter from Swedenborg to Benzelius gives a glimpse of the lively international interactions that now took place at the Bureau and at Uppsala: "Reaumur, who has written about steel in France, has come here, and is at the [Bureau]; from what I have seen of him already, I consider him to be a clever scientific man. . . . When you answer Dr. Sloane in England, you may mention that we have spoken on the subject together; and that I am willing to correspond with them on matters concerning metallurgy, if they are willing to print it at their own expense."[64]

The fledgling attempts to create a thriving milieu for knowledge production around the Bureau of Mines and Uppsala University had finally met with success. The Cartesian circles at Uppsala University had connected to the Bureau of Mines, and an enduring common ground had been established between the two. Curious knowledge had been rejected and replaced with a fascination with ingenious inventions, which eventually would cede to a general agreement that natural philosophy, above all, should be useful.

Chymistry, however, had not joined itself to the new movement. We have seen in the previous chapter how the aging Hiärne continued his experiments in transmutative chymistry and continued to defend a basically Paracelsian position. Chymistry remained something of an anomaly for the first generations of mechanical and experimental philosophers at the Bureau. None of them had any deeper knowledge about it. Doubtless these men had a degree of respect for chymistry's knowledge and the methods of the chymists themselves, but they could not come to grips with it on mechanical and geometrical terms.[65] Simultaneously, they discussed gold making with open hostility.[66] In two manuscripts from the 1720s, Polhem explained why artificial gold making was impossible. He agreed that minerals and metals were created through an act of transmutation, which took place deep inside the earth. This was a standard view of the origin of metals.[67] Then he gave the theory a mechanist twist when he argued that the creation of dense materials required very strong mechanical compression. Deep inside the earth, this force could be created by the overlying layers of rock, but it was impossible to create such a force by human artifice on the earth's surface. Therefore, it was impossible to create gold out of less dense metals.[68] Polhem's discussion was a typical speculative mechanical hypothesis in the Cartesian vein, which could not be verified, and which left open the chymical issue of how the metals actually were made.[69] Polhem also admitted in private that he was a poor chymist and that he did not have sufficient knowledge of chymistry to give a complete description of the processes that shaped the earth and the metals.[70]

Swedenborg studied chymical books in earnest, and made systematic comparisons between his new study and what he already knew of geometry and mechanics. However, he never took the step into laboratory practice.[71] His written output in this field consisted mainly of reviews and summaries of literature. In his monumental work on iron, *De Ferro* (1734), Swedenborg made a rhetorical point of distancing himself from secret mongers, alchemists, and artisans. These groups, claimed Swedenborg, did not want to partake in the free exchange of knowledge and would keep what they knew as private secrets. The real lovers of knowledge, however, laid it out bare before the public.[72] By criticizing the incomprehensibility of alchemy, Swedenborg turned his ignorance of chymistry into a virtue.[73] Because the main part of the corpus of the chymical literature was dismissed, there was no need for any chymical theories. Actually, *De Ferro* is a predominantly utilitarian work. For example, Swedenborg related several methods of how to turn iron into crocus martis, which was used in glass manufacture. He discussed how red pigment could be made from iron vitriol, and how iron was used in various medicines.[74]

Polhem's and Swedenborg's theoretical discussions were kept within the safe framework of physics and mechanical philosophy, as when a mechanical theory of atoms was used in a discussion of magnetism. One of the few longer theoretical sections was based on Robert Boyle and discussed how iron could increase in weight. In comparison, the discussion of the growth of iron ore in the earth, in slag heaps, and in water was kept purely descriptive and devoid of a theoretical framework. If such a framework had been developed, it would necessarily have been a chymical one.[75]

Here was a comprehensive critique of chymistry and an attempt to colonize chymistry with speculative mechanical philosophy. Simultaneously, chymical theories and theoreticians were given the silent treatment, and theories of transmutation were denigrated as senseless and useless speculation connected to alchemical gold making. As we have seen, however, Swedenborg's theories did not do much to explain chemical phenomena. Therefore this type of critique of chymistry could never be particularly effective, unless it was also embraced by the chymists themselves. It had little effect when pitched, as it were, from the outside by individuals with little knowledge of chymistry. But soon there would be a new generation of chymists about, who had internalized mechanical and experimental philosophy, and who would make arguments such as these their own.

Mechanizing Chymistry: The Boerhaavian Tradition

In the first decades of the eighteenth century, there was still no formal education to be had in chymistry at Uppsala University. The university did, however, turn out a good number of promising mathematicians and mechanical philosophers. Geometry had a central position in their thinking. As geometry expressed absolute, irrefutable truths, many aimed to describe the world in its entirety as a geometrical system. This style of thought emphasized order, simplicity, and harmony and sought to abstract geometrical shapes from nature. Through such operations, the world became a geometrical machine in constant motion. It was by use of mechanical metaphors that both nature and thinking were mechanized, and reconceptualized as analogous to human-made machinery.[76] Among this group of young philosophers was Georg Brandt, a student of the Uppsala mathematician Anders Gabriel Duhre.

Brandt's formal education at Uppsala did not comprise the full extent of his knowledge. He was also the son of a Stockholm apothecary who had sold his pharmacy and bought an ironworks in central Sweden. In Brandt thus is the conjunction of pharmaceutical and mining skills, which Hiärne had deemed ideal for the well-educated Bureau chymist.[77] In 1718, Brandt edited and published Duhre's mathematical lectures, in which, surprisingly, Brandt wrote an introduction containing a long discussion of chymistry. The piece had been written during his time as an auscultator at the Bureau of Mines, and it was dedicated to its president and governing board. Therefore the book, and its introduction in particular, should be interpreted as an appeal for patronage through a display of skill and a statement of scientific positions and intent.[78]

Brandt wanted to put chymistry on a mathematical and geometrical footing. The example to be emulated was that of Newton. Indeed, the correct application of mathematics to experience was, according to the young Brandt, absolutely necessary if one were to arrive at correct knowledge about nature.[79] After praising Newton, Brandt proceeded to explain how mathematics could serve also as a foundation for chymistry. His point of departure was the Englishman "Johan" (John) Freind's Oxford lectures of 1710.[80]

Brandt began by agreeing with Freind's criticism of earlier chymical authors, whose writings were described as nearly impossible to understand.[81] Both their fundamentals and the language that expressed them were obscure. Given this, one could logically conclude that what was said was empty of content. Put a different way, Freind, and Brandt, claimed that to have a proper foundation, chymistry needed to proceed from axioms in a manner similar to Euclidian geometry. That is to say, when the axioms were unknown, there was no foundation on which to build a theory. Then a new theme was introduced. It stated that the authors of older chymical works must either be ignorant or know what they write about but deliberately choose to hide it from their readers. Both propositions served equally well to discredit the chymical literature, for either it was written by ignorant persons, or it failed in its proper task to give the reader access to the knowledge of its authors: "hence it may not seem strange if they have built nothing on a foundation of nothing; because they pronounce themselves, he says, as if their intent were to hide their own ignorance, or at least as if they would not like to give others enlightenment in this matter."

Brandt hammered home and expanded this point in the following passage. He disassociated books on chymistry both from the study of nature and from the authors themselves. These books did not agree with nature, and they served no purpose, even to further the aims of their own authors! The foregone conclusion from the discrediting of the whole of chymical literature was that chymistry lacked a solid foundation, and therefore that it was a pseudo-knowledge in its present state. It had no axioms to serve as solid building blocks for theories. Chymistry had been on the wrong track all along, as it had not investigated "the mechanical nature of bodies." How these investigations were to be pursued in practice was, however, not specified. After this broadside, Brandt presented a more precise critique of contemporary chymistry. The problem he saw seems to have been the theory that there were layers of proximate principles, which could be arrived at as physical substances through laboratory analysis. Brandt associated it with wishful thinking: "but they have rather fabricated such as one as they would have wanted it to be, and therefore have ascribed bodies the layers and properties that opposed mechanics as well as themselves." Finally, Brandt approved of

Freind's replacement of such theories with demonstrations of geometry and physics.

Compare Brandt's position to that of Odhelius as outlined in the previous chapter. Odhelius disapproved of chymical theory in general, and what he had come to regard as the writings of the alchemists in particular. He had, however, been equally disdainful of rationalist physics. Hence, he ended up on a similar position to that of Boyle's skeptic in the famous *Sceptical Chymist*: the chymical theories were no good, but there were also no viable alternatives. Brandt, on the other hand, attempted to colonize chymistry with geometrical demonstration. But these attempts could only go so far. They were excellent in providing a logical and structured criticism of the style of older chymical authors, but they were, as the experienced Odhelius had observed in another context, unconnected to the chymical practice of the laboratory.

Brandt's style of argument, however, did not present him at a disadvantage at the governing board of the Bureau of Mines. The Bureau had come to assign a high value to mechanical and mathematical knowledge and had grown skeptical of transmutative chymistry. In 1721, a few years after publication of the book, Brandt was given a traveling grant, and he immediately set off to Leiden to study chymistry and physics. Apparently, it was not self-evident that a foreign higher education in medicine and chymistry was the proper way to spend the Bureau's chymical stipend. When he applied for the grant, he claimed that his intention was to travel to foreign mining installations. In his application for permission to travel abroad, he specified that he would visit foreign places and mining sites.[82] Later, in a slightly cheeky 1723 letter from Leiden, Brandt apologized that he had not visited any mining installations, but had stayed on in Leiden "for the sake of chymistry and physics." He would stay for the winter too, to further his knowledge in these subjects. God willing, he would visit European mining sites before returning home.[83]

The Bureau forgave him. The University of Leiden was doubtless Europe's most admired site for the teaching of medicine and chymistry in the decades around 1700. It drew students from all over Europe. Studies in Leiden and a subsequent Dutch degree in medicine were symbols of learning and status that were recognized everywhere.[84] This was not lost on the keen satiric eye of Jonathan Swift. In the foreword to his *Gulliver's Travels* (1726) Swift endorsed the credibility of his protagonist with the claim that not only was Gulliver known by his neighbors to be the most credible man of his native district but he had also studied in Leiden.[85]

Beginning in the 1670s, the teaching at Leiden's medical faculty had increasingly been oriented toward England and the research program promoted by the Royal Society. In the hands of the professor of chymistry, Herman Boerhaave, this program gained a chymical and utilitarian interpretation.[86] Boerhaavian *chemistry* (as we may call his endeavors) was explicitly devoid of nonmaterial aspects. The aim of chemistry was to unite and separate substances, and to study mechanical and chemical change in material objects. As John C. Powers has observed, Boerhaave's "course provided a model for practicing chemistry as a form of experimental philosophy."[87] Boerhaave attributed his method for constructing structured experiments to Francis Bacon, but the method's application to chymistry was wholly his own. In Boerhaave's vision, mechanics, physics, and chemistry were complementary. Chemistry was both a useful art and an integrated part of the new natural philosophy.[88]

According to Boerhaave, the many errors that could be found in earlier chymical works were the result of "excessive and undisciplined theorizing," along with a lack of a proper methodology. Belief in gold making fell into this category of erroneous thinking, as well as Paracelsus's discussion of elemental and monstrous beings. The latter was dismissed as a promulgation of the superstitions of the uneducated. In Sweden as elsewhere, Boerhaave's method would serve as a methodological template for the ordering of chemical facts, and for the evaluating and making of chemical knowledge. Boerhaave's methods provided his students with a working model for how to conduct chemical experiments, but it also "divided, defined, and classified chemical entities in specific ways, which underwrote his natural philosophical views."[89] This latter aspect of Boerhaave's chemistry connected strongly to the knowledge area of natural history and would not fall on deaf ears among his many Swedish adherents. Here was a chemical theory and a method that fitted well with the ideas that were proposed among mechanists, physicists, astronomers, and mathematicians. In addition, Boerhaave combined his theories with a full-blown experimental research program, which made use of a rich set of chemical investigative methods. Undoubtedly, Brandt had arrived at the right place.

The Laboratorium Chymicum of the Bureau had lain dormant for most of the 1720s. Little work had been conducted there during the final years of Hiärne's custodianship. His successor, Magnus von Bromell, had not made any particular efforts to revive it. After his return from his journeys, Brandt was informally given the responsibility of the laboratory in 1727 and was made guardian of the Mint in 1730.[90] He had been appointed to lead the reconstruction of the Laboratorium. The

old building was sold, and the laboratory was relocated to the Royal
Mint. The Bureau's main office was located in the same building, and
hence the move tied the laboratory tighter to the administration.[91] As
guardian of the Mint, Brandt enjoyed free board at his place of work, in
addition to a room used as laboratory, and another used as a chamber
of assaying.[92] Brandt was not to be permitted the independence that
Hiärne and Bromell had enjoyed. He was provided with lodgings and
laboratories, but was also kept under closer scrutiny by the Bureau's
governing board. Furthermore, Bromell continued to be the official head
of the laboratory until his death in 1731, and Brandt was in fact not
promoted to an assessor until 1747, and became the official leader of the
laboratory as late as in 1748.[93] But Brandt's slow advance in the ranks
of the Bureau should not be taken as an indication that the Bureau disapproved of him, or his work. He had in fact a rather typical career. It
was Hiärne's and Bromell's careers that had been exceptional, insofar as
they had been permitted to skip ahead in the Bureau's hierarchy.

Brandt's views on transmutative chymistry—or alchemy, as it was
now routinely called—have been a matter of discussion. Torbern Bergman, in his oration to the memory of Brandt at the Swedish Royal Academy of Sciences, claimed that "Brandt was not completely free from
Alchemical trials the last years of his life; but He set them up and completed them from another foundation and urge, than the so called Gold
makers."[94] Axel Fredrik Cronstedt, too, made a similar remark.[95] There
are also no extant published texts in which Brandt criticized transmutation of metals. A different picture emerges from a set of notes catalogued
at the Uppsala university library as Brandt's chemical lectures. In this
semi-private context, Brandt was disdainful of seekers of the philosophers' stone, and attempted to disprove older theories of matter, and
Paracelsian matter theory in particular.[96]

It is likely that Torbern Bergman was on the right track. Brandt may
have conducted experiments in metallic transmutation as a Boerhaavian
exponent of Bacon's notion of experimental history. That is, he may
have sought to collect and repeat the experiments and operations of
older chymists. As shown by Ursula Klein, Boerhaave conducted such
experiments with the aim of translating the language of chymistry into
the mechanist discourse.[97]

The greatest part of Brandt's work, however, went into the careful
chymical analysis of minerals. He began to publish his investigations in
the journal of the Uppsala society of science. The society and the journal
were the immediate heirs of the College of the Curious, and of Swedenborg's journal *Daedalus Hyperboreus*. His first chemical publications

consisted of investigations of gold and mercury (1731), and another on arsenic compounds and alloys (1733), which carefully outlined their characteristic differences and relationships with each other.[98] Then followed his most famous publication, *Dissertatio de Semi-Metallis* (1735), containing his analysis of cobalt and characterization if it as an independent semi-metal.

It fell to Brandt to reestablish the Laboratorium Chymicum as a site for advanced research, and to take on the role of teacher for a new generation of mining chemists at the Bureau. According to the formal instruction to the laboratory, its leader should teach four auscultators each year in "the knowledge of ores, chemistry and assaying, as well as the application of these sciences to smelting."[99] Hence quite a few of the Bureau's future officials passed through Brandt's laboratory. Simultaneously the relatively small number of students at any given time ensured that teaching proceeded on a small scale in a close and personal relationship similar to the way artisans transferred knowledge from master to apprentice.

Brandt would become the grand old man of a circle of chemists who defined their art as emerging from mechanical philosophy. These men were trained in a chemistry and assaying that was geared to cater to the needs of the mining business. Other participants in this circle were Henric Teophil Scheffer (the Bureau's head assayer, 1740–59), Axel Fredrik Cronstedt, Sven Rinman, Benct Qvist, and Cronstedt's protégé Gustav von Engeström, who would become Brandt's successor as director of the Laboratorium Chymicum.

In the following years these men built a strong internal tradition of chemistry at the Bureau. The group shared the common view that chemistry, together with mathematics, mechanics, natural history, and some other topics, comprised the foundation for practical applications in mineralogy, metallurgy, and assaying. In practice they were preoccupied with mineral analysis, metallurgical trials, smelting trials, mineral systematics, and with the teaching of these subjects. The officials of the Bureau held what they had come to regard as their scientific knowledge in very high regard. Chemistry, defined as a practical and useful enterprise cut lose from alchemical speculation, was a vital part of it.[100]

Boerhaavian chemistry struck a resonant note not only with Brandt and his students. When Johan Gottschalk Wallerius was made Uppsala's (and Sweden's) first professor of chemistry in 1750, he based his courses and his main textbook on Boerhaave's *Elementa Chemiae* of 1732. In a draft of biographies of chemists, Wallerius's successor Torbern Bergman wrote about Boerhaave that he was the "first to unite Chemistry with true physics." In contrast, Bergman claimed of Georg Ernst Stahl that he "wrote in almost all parts of medicine and Chemistry, but darkly. [He] won much praise, but also much criticism." The teaching of Boerhaave and this textbook can thus be regarded as the ur-text, or original source, for the chemistry conducted both at the Bureau and in Uppsala.[101] The recruitment patterns at the eighteenth-century Bureau meant that almost all officials where Uppsala alumni who subsequently received specialized training at the Bureau. As a consequence the new generation of chemists at the Bureau came to be almost completely disconnected from Paracelsian philosophy, and from transmutative chymistry.

: : :

The introduction of mechanical and experimental philosophy at the Bureau followed that of chymistry with a delay of approximately two decades. Through the journeys of Odhelius, Swedenborg, and others, the mining officials at the Bureau gained a thorough acquaintance with the new experimental philosophy. It was introduced simultaneously at Uppsala and the Bureau, through the concerted effort of the circle around the Collegium Curiosorum. The same group would found a new journal and generally promote the importance of practical and useful knowledge. The common denominator of the Bureau and the Uppsala group was utility. This word, *nytta*, would become a keyword that would bring together Swedish elites around natural philosophy. At the Bureau there were also further components: natural history and the reinterpretation of chymistry as mechanical chemistry, as outlined by Boerhaave. All of these components would be integrated into the chemistry of Georg Brandt. Brandt would teach mechanical chemistry to the future state officials, effectively severing them from the previous Paracelsian and chymical traditions current at the Bureau. Building on Boerhaave and Brandt, subsequent officials, and chemists, at the Bureau would oriented themselves strongly toward the new experimental philosophy, as developed by natural philosophers in England and the Netherlands.

This is not to say that there was no, or little, communication going on between this group and chemists and natural philosophers in areas of Europe outside of England and the Netherlands. There was, for example, an ongoing translation of Swedish-language mining and chemical literature into French.[102] Direct exchanges, however, seem to have been limited mostly to the area of iron and steel.[103] As mentioned earlier, Réaumur visited Sweden and the Bureau of Mines. Swedenborg would make extensive use of his *L'art de convertir le fer forge en acier* (1722) in his own *De Ferro*.[104] But Réaumur left few traces in the subsequent works of the chemists and mineralogists at the Bureau. Like Swedenborg and Polhem, Réaumur was interested in the microstructure of materials: their choice of problems emerged from physics and concerned the structure of matter.[105] The mechanical chemists of the Bureau, on the other hand, were concerned mostly with chemical composition. It was in the latter area, as well as in natural history, that they expended most of their efforts and would make the greatest impact. What then of the Holy Roman Empire? As I show in the following chapter, there were intense interactions going on between the Bureau's officials and their counterparts in the mining regions of central Europe.

5

Elements of Enlightenment

In Sven Rinman's *Bergwerks Lexicon*, a 1788–89 encyclopedia of mining arts and mining sciences, the entry
on cobalt covered several pages. Rinman explained that
cobalt was an ore containing a quite recently discovered
semi-metal, "cobalt-metal." Some varieties of the ore
consisted of this metal and arsenic, iron, and/or sulfur,
and others of cobalt-metal, iron, and sulfuric acid. Basing
his entry on the latest German- and Swedish-language research in chemistry and mineralogy, Rinman carefully described how the various varieties of the ore looked and in
which locations they could be found. He also delineated
the chemical properties of the ore and the metal, and their
practical use in manufacture. At the end of the entry, almost as an afterthought, he added the following passage:
"*Kobolt*, or *Cobolten*, have also several other things been
called at the German mine works in olden times, such as
firstly their imagined little ghosts, *Gnomes* or *Bergmän-
lein* in mines, secondly poisonous fumes, or *Schwaden*,
thirdly white arsenic-bearing kies, or misspickel. Pure arsenic, or *scherbenkobolt* and other substances have also
been given the name of cobalt."[1]

In this simple and straightforward way, Rinman dismissed not only the existence of the kobolder, but he also
dismissed the miners who believed in them. Maybe the
connections between ghostly underground denizens, poi-

sonous fumes, and enchanted ore were lost on him. In any case, he ignored them. To Rinman, kobolder were simply figments of imagination,
and cobalt-metal was simply a fact of nature. Just like, for example,
platinum, silver, lead, and zinc, it was both a stable, marketable commodity and one of the basic components that made up the material
world. His description proceeded from what must be seen as the most
important innovation of the Bureau chemists: the notion that metals
were basic species of the mineral realm and should be considered as
building blocks from which the world of matter was composed.[2]

In the previous chapter we saw how chymistry was turned into mechanical chemistry. The present chapter investigates how mechanical
chemistry was combined with assaying and natural history to engender
mineralogical chemistry, a new field of investigation based on the conception that the material world was comprehensible—in its entirety—in
terms of chemical composition. This conceptual framework supplanted
other theories and discourses about the generation of metals, and consequently permitted Rinman and other eighteenth-century mining chemists to disregard both mine gnomes and metallic transmutation. Their
new vision of mining chemistry had no place for kobolder or subtle
matters that could not be translated into monetary value, or reduced to
their component parts in an industrial process or in a laboratory. There
was also no room for miraculous transformations, only for the rear-
rangement of discreet parts.

The centerpiece and organizing principle of this new mineralogical
chemistry was the identification of metals as basic species of nature.
Or to put this concept in twenty-first-century language, the proposition
that metals are to be considered chemical elements. The development
of this notion engendered a prosperous and highly influential research
program at the Bureau. Indeed, as observed by Colin Russel, "as late as
1886 no less than 40 per cent of the chemical elements found since the
Middle Ages had been discovered in Sweden."[3] Although this research
tradition began at the Bureau, what interests me here is not to establish
the priority or importance of the discoveries of the Bureau chemists. It
is to investigate the natural philosophical and cultural foundations, and
consequences, of their research program. The idea of metallic chemical elements was an innovation constituting nothing less than the coming into being of a new ontological category: a new general kind that
was used to organize the world of experience.[4] It proceeded from mechanical philosophy, and its conception of material reality as a machine
composed of discreet, identifiable parts that could be combined and
recombined. In the hands of the Bureau chemists and mineralogists, this

philosophy received a new empirical and economic foundation. The metallic building blocks of nature that they proposed were not only solid and tangible; they were also commodities that could be sold and bought on the market.

The importance of cameralism for this epistemological position cannot be emphasized enough. Cameralism was a complex of economic theories and practices that served to guide state policy in several northern and central European seventeenth- and eighteenth-century states.[5] It was essentially rationalist and authoritarian. In the words of Lisbet Koerner, cameralists "hoped to draw together the state's various domains into one administrative unit, and to regulate that state's material links with other polities. To that end, they employed many tools: producer monopolies and state manufactures; import barriers; navigation acts; pro-natalist legislation; export bans on gold; funding for science and technology; and improvements to infrastructure."[6] Cameralist thought had a strong foothold and was partially developed in the mining administrations of the European north and the Holy Roman Empire. The *Direktionssystem* (Ger.) favored by continental European mining states tied domestic mining enterprises strongly to the state. It sought total control over all aspects of domestic production.[7] Artisanal autonomy was to be reduced, crafts were to be inspected, controlled, and reformed, and workers were to be forced to become cogs in the great, centrally planned, machine of society.

To increase control over, and hence revenues from, the mining business it was necessary that state mining officials had a deep and intimate knowledge of the various artisanal processes used to excavate and process minerals. From their point of view, mechanics, surveying, assaying, law, and other "mining sciences" constituted the knowledge of the various parts of an integrated system of production.[8] This system required detailed governance, or policing. Mining was the proper arrangement of resources. Metal-bearing ores, charcoal, building wood, machinery (and streaming water to power it), as well as skilled artisans and laborers: all of these elements were to be collected and arranged in a proper sequence. When all parts of this machine for metal production were correctly assembled, metal, the valuable end product, would be produced. The role of the mining sciences in this system was to optimize the process/machine of mining. Mining officials who framed their knowledge in terms of mining science rather than craft knowledge also legitimized their status as overseers and custodians of the entire production system. Systematic and comprehensible—scientific—mining knowledge was deemed to increase productivity, and therefore also state revenues. As

FIGURE 7. A Danish mechanical cabinet, made about 1750, showing the mechanized cameralist world-view in action. Mining is a machine, with the king (Frederic V of Denmark) at the top and the miners at the bottom. Crushing, smelting, mining administration, and minting are in between. The cabinet can be folded in the middle to form a closed box, and the figures move when a crank is turned. Reproduction by Tor Aas-Haug, Norwegian Mining Museum, Kongsberg.

Wakefield has put it, "the cameral sciences were intimately and ineluctably tied to the sciences of nature, especially chemistry and mineralogy."[9] Consequently, a number of theoretical and methodological developments in chemistry and mineralogy were elaborated in interaction with the transformation of these knowledge areas into cameral sciences at the service of Europe's elites. The elaboration, during the first half of the eighteenth century, of the notion that metals were basic species of nature was one of them.

Nowadays the work of the many chemists who were active during the first half of the eighteenth century is largely forgotten. Research on Swedish chemistry and mineralogy, for example, has tended to focus on the period from about 1760, as well as on Torbern Bergman and his associate Carl Wilhelm Scheele, and on Jöns Jacob Berzelius, active from about 1800. The latter-day fame of these men has to some degree overshadowed that of their Bureau predecessors and associates, and that of Johan Gottschalk Wallerius, Bergman's predecessor as professor of chemistry in Uppsala. This, I think, is unfortunate. As we will see, the later generation drew extensively on the work and experience of their predecessors, as well as on the work of Carl Linnaeus, whose active period coincided closely with that of Wallerius and the Bureau chymists.

It is also necessary to realize that the concept of "Swedish chemists" is misguided. Indeed, the Bureau chemists and mineralogists of the period studied here, ca. 1730–60, were as deeply enmeshed in Europe-wide networks of knowledge circulation as ever before, or after. To bring home this point, and to situate the Bureau's place in the wider context of European mining and smelting, this chapter begins with an outline of the Bureau's place in the wider context of central European technical and chemical reforms of mining. Then it moves on to the laboratories and chambers of assaying of the Bureau itself and discusses the efforts of Brandt, Cronstedt, and other officials to pursue economic and industrial reform by reforming chemistry and mineralogy.

The Bureau of Mines in the Context of Central European Mining

From medieval times and throughout the sixteenth and seventeenth centuries, the core of European mining and smelting know-how was situated in the German-speaking regions of central Europe, extending into Italy and Hungary in the south, and Sweden and Norway in the north. Knowledgeable artisans from main mining centers, in particular Saxony, Austria, and the Oberharz area, were readily sought after and traveled widely. The area that would become the main charge of the Bureau's officials—the Bergslagen mining district of central Sweden—was a northern extension of a Europe-wide circuit of artisanal migration and knowledge exchange. Immigrant Germans had established new mines and smelting furnaces there already during the Middle Ages.[10] In the seventeenth century, Swedish territorial expansion into German-speaking areas had further facilitated communication and also active recruitment. In a sense, mining knowledge was a spoil of war. The relative ease with which the Swedish state could recruit during this period may have been a consequence of the devastation of many of the mining areas of central Europe during the Thirty Years' War.[11]

Although workers and artisans from Sweden were employed at continental mining sites, the country tended to be on the receiving end of exchanges up to the beginning of the eighteenth century. The seventeenth-century expansion of the Swedish mining enterprise was to a large extent dependent on immigrant knowledge-carriers, the majority of whom were German speakers.[12] In sheer scale of immigration, they were rivaled only by the 2,000 Calvinist Walloons (mainly forge men, smelters, and their families) who emigrated to Sweden from the Liège region in the 1620s and 1630s.[13] Many Germans were skilled artisans of middling or fairly high social status, such as master builders, assayers, smelters, and mining administrators. Foreign artisans and administrators were particularly active at important sites, such as the mines of Falun and Sala.[14] Even the Swedish mining terminology was essentially German. Rinman's *Bergwerks Lexicon* (1788–89), of 1,248 pages, concluded with a 24-page register of German terms used in the main text.[15]

The Bureau of Mines too, had mainly built up its knowledge through the recruitment of foreigners. One of the driving figures behind the foundation of the Bureau in the 1630s had been the assessor Georg Grissbach. Born in the Harz region, Grissbach came from a family of mining administrators. His grandfather had been master of the mine at Sankt Andreasberg. Other influential immigrant officials were the assessor David Friedrich von Siegroth and the scribe Jost Franck. Recruitment to the highest offices soon came to a stop, however. Later in the seventeenth century, leading positions were often held by Swedish aristocrats, whereas foreigners received mostly lower offices. This arrangement allowed the aristocracy to maintain social distance from those beneath them in the state hierarchy, while simultaneously permitting a massive influx of foreign artisans.[16]

A few decades later, lower offices too would be closed to foreigners. From the beginning of the eighteenth century onward, most officials were recruited from among the local elites of the Swedish mining districts. They were mostly sons of ironmasters, mine owners, and state officials. They were also, to a substantial degree, the sons or grandsons of immigrants, and as discussed in the previous chapter, they were mostly Uppsala alumni. This new breed of mining officials took on a more assertive and active role in exchanges with their continental and English counterparts.

They also fostered patriotic sentiments at the Bureau. In the eighteenth century, Sweden's elites were slowly coming around to the insight that theirs was a young nation when it came to culture and knowledge. The grand historical theories of the seventeenth century that had declared the country to be the original well-spring of human culture and the mother of all nations were now treated mostly with irony.[17] With the realization that local cultural expressions were highly dependent on foreign influences came a competitive will to defend what were regarded as recent Swedish accomplishments. "Swedish" chemists and mineralogists were beginning to differentiate themselves from their counterparts in other nations and to present their work as patriotic achievements. In particular natural history was singled out as a specific area of accomplishment, along with cameralist economy, chemistry, and the mining sciences.

There was indeed some prowess. The period of circa 1730–80 has been described as the heyday of Swedish science. During this period the Bureau of Mines as well as Uppsala university gained wide recognition as important sites of knowledge production. These places became objects of admiration and destinations for travelers. Toward the middle of the eighteenth century, European actors around mining, chemistry, and mineralogy were acutely aware that something interesting was going on in "Swedish mining." This awareness was not always formulated in a positive way. Perhaps feeling the heat of the competition, the Göttingen chemist, mineralogist, and chief police commissioner Johann Gottlob von Justi discussed the contribution of Swedish mineralogists in the foreword to his *Grundriss des gesamten Mineralreiches* (1757). Justi praised the knowledge of the Germans and denigrated the immediate predecessor of his work, the influential *Mineralogia* of Wallerius. It would not do, Justi stated, that the Germans who had taught the other nations the arts of mining should be content with "the translated textbook of a foreigner." Besides, Wallerius's work was riddled with errors. Justi was also dismissive of the chemical and mineralogical works of the Bureau chemists.[18]

The widely disseminated review journal *Göttingische Anzeigen von gelehrten Sachen* took a more moderate stance. Justi was rebuffed for provoking the Swedes, but *Anzeigen* too held that the Germans had taught mining to the other nations. The Swedes had but recently risen to prominence in the area and should acknowledge their debt to their teachers.[19] A similar point was made more bluntly in a report from Sweden, written in 1747 by the Hanoverian mining administrator Johann Carl Hansen for the benefit of his superiors in Oberharz: "The owners of the ironworks are deceitful, just as if they were artisans, when they notice that foreigners approach. They are of the opinion that Sweden is the only place in the world where one understands how to make iron. Although it is not long ago that the stupid Swedes sent their iron to [be refined in] Germany."[20]

Hansen was clearly irritated. Not even bribes gave him access to the smelting process, and he had to be content with watching it in secret, at night. Judging from these grudging voices it would seem that "German" dominance in mining knowledge was coming to an end. But was it really? Actors' articulation of the lines of national division—their patriotic boundary work—must be distinguished from their practice. Early modern states were highly competitive entities. We should not be surprised that Swedish state officials sought to cast themselves as successful and diligent examples that should be emulated. We should also not be surprised that men like Justi and Hansen, who sought patronage and positions in other states, considered it a viable strategy to denigrate the "Swedes." In practice, however, the knowledge of natural philosophers and mining officials in Sweden and the Holy Roman Empire was highly interdependent.

One of a few well-studied examples of interactions illustrates this point. It concerns Christopher Polhem's several attempts to improve the system of mine pumps in the Harz.[21] Polhem traveled to the Harz mining region, and in 1747 a Hanoverian delegation visited Stockholm. Among the visiting officials was a young Friedrich Anton von Heynitz, who later would become the most famous of all reformers of the mining administrations in the Holy Roman Empire. Polhem was paid by Hanover to train, in all, four students. Three of them, Bernt Ripking, Christian Schwarzkopf, and the above-quoted Johann Carl Hansen, later became high-ranking mining officials in the Harz.[22] According to Bartels, Polhem's training of these men was something of a starting point for a general change in the use of mining technology, and for cartographic practice in the area.[23] To what extent there was a thoroughgoing Swedish influence over German—that is, Oberharzian—mining technology is, however, a moot point. Polhem was, after all, the son of German immigrants and worked in an organization that constantly looked to the Oberharz and Saxony for inspiration. What is significant here is rather the intensity of exchanges and the realization of the eighteenth-century Oberharzian mining administrators that knowledge exchanges with their corresponding Swedish officials went both ways.

The situation was similar in the fields of chemistry and mineralogy. Differences tended to be articulated on theoretical rather than patriotic grounds, and the nationality of individual scholars was often left out of the argument. In the second part of his *Lithogeognosie* (1751), a work on the chemical analysis of earth and stones, the Berlin-based author Johann Heinrich Pott noted that he had always given due credit to his worthy predecessors Hiärne, von Bromell, and Henckel. Then he went on to criticize the works of Johann Lucas Woltersdorf and Linnaeus. Not once did he mention that Henckel and Woltersdorf were Germans and the other three men Swedes.[24] Similarly, Cronstedt explicitly mentioned Pott as a main methodological inspiration for his mineralogy of 1758.[25] In a

1760 pamphlet written as an answer to von Justi's criticism of his work, Pott proclaimed that although Justi claimed to have found fault with him, Cronstedt had found his experiments to be correct. Pott continued by praising Cronstedt's mineralogy, which in his view had been written with great skill, judgment, and experience.[26] The work was translated into German in 1770 by a Dane: Morten Thrane Brünnich, professor of cameralist economy and natural history at the University of Copenhagen. A new German translation appeared in 1780, conducted by the Freiberg mineralogist Abraham Gottlob Werner. Werner presented it in a laudatory preface that emphasized the novelty of the system and its great contribution to mineralogy as a whole.[27]

As shown by these examples, knowledge production at the Bureau and elsewhere depended on circulation of knowledge across state boundaries. The chemists and mineralogists who were active in Saxony, Hanover/Harz, and Austria, the main mining states of the Holy Roman Empire, as well as Berlin-based chemists such as Pott, comprised the most important group the Bureau's officials related to: in writing, through travel, and through correspondence.

This insight, however, has not informed historical scholarship. Studies of the various mining enterprises of the Holy Roman Empire have tended to overlook the importance of exchanges with Sweden.[28] Wolfhard Weber's magisterial biography of Heynitz presents an impressive wealth of information about Swedish-German contacts, about Heynitz's journey to Sweden, as well as about the influence of the organizational model provided by the Bureau of Mines on his efforts at reform. In his conclusion, however, Weber stresses only the influence on Heynitz of the French and English examples.[29] Similarly Baumgärtel does not take any particular note of Swedish-Saxon interactions, but quotes influences from England, Hungary and the Harz region as important to the Freiberg mining school.[30] German historiography is well matched by a similar Swedish lack of interest in German developments. Lindroth, for example, stated that in the area of mining mechanics "we [Swedes] had just about nothing to learn from foreign lands" in the late seventeenth century.[31]

Nevertheless, state officials in Sweden and in the mining regions of the Holy Roman Empire shared a common culture. An essential feature of this common culture was that it was defined in cameralist terms. Cameralist officials—regardless of which state they served—were interested in similar issues of governance and faced similar problems. It was this common ground that made admiration, inspiration, as well as criticism and denunciations of the other's pride of accomplishment possible.

FIGURE 8. A Freiberg furnace for the purification of silver with lead. An example of the great number of drawings and descriptions of mines and mining installations in the Holy Roman Empire that can be found in manuscript collections associated with the Bureau of Mines (variations of this drawing can be found in several of them). From the notebooks of A. F. Cronstedt, call number F1:13, p. 33, KVA. © Center for History of Sciences, the Royal Swedish Academy of Sciences.

In the eyes of its eighteenth-century contemporaries, the Bureau's prominence derived to a large part from its perceived ability to reconceptualize the various knowledges used in mining and smelting as cameralist mining sciences. The alleged Swedish ability to transform the knowledge of other nations into fully fledged systematic knowledge, or science, was noted as a particular source of patriotic pride. Axel Fredrik Cronstedt, a young official (*Geschworner*) at the Bureau, made a revealing remark to that effect. It was made toward the end of the speech he gave on his 1754 introduction as a fellow of the Swedish Royal Academy of Sciences: "It seems as if it falls to the Swedes, to bring to conclusion the work of foreign learned men, with the help of the observations that they contribute."[32]

Intelligence gathering held, as we have seen, a central role among the Bureau's activities, and here it was presented with a patriotic edge. Cronstedt cast Sweden as a center of knowledge, and the Swedes as especially suitable to synthesize and put to order observations that had been made elsewhere. He was of course also aware of the European impact of Linnaeus's botany and zoology. The remark can be interpreted as veiled praise of Linnaeus, one of the founders of the Academy. But it is more likely that Cronstedt was thinking about his own program of discovery of metals paired with mineralogical reform. It was a program that derived both from the analytic approach of the Bureau's chemists and assayers, and from the natural history traditions of Linnaeus's Uppsala.

Discovering Metals, 1730s to 1750s

Of old, Western matter theory acknowledged the existence of seven metals. Their names were gold, silver, copper, iron, tin, lead, and mercury. There were many theories concerning the creation of metals. Some held to the traditional medieval chymical position that they were composed of primal mercury and sulfur. Others proceeded from the Paracelsian tria prima of salt, mercury, and sulfur, and some, like Jan Baptista van Helmont, proposed that all substances in the world were made from, and could ultimately be reduced to water.[33] Regardless of their leanings, most chymists and other matter theorists were in agreement that metals were composite substances that were formed underground through the combination of simpler ingredients, or through a process of growth, and that they were divisible into their constituent parts through chymical analysis.

Toward the end of the sixteenth century, accumulated experience of mining and assaying, as well as chymical experiments, had established the existence of a number of metal-like substances. These were the so-called half metals or semi-metals. Like the true metals, they were composite substances, but unlike them, they had failed to come into a wholeness of being. As Paracelsian matter theory would have it, they were originally intended by nature to mature into whole metals, but failed and only reached a partial stage of development. The reason was that they had been unevenly fed with salt, sulfur, or moisture (mercury) or had been hindered in their development for some other reason. Hiärne regarded metallic mercury, as opposed to philosophical mercury, that is, the principle of moisture, as a semi-metal. Others were antimony, bismuth, cobalt, arsenic (of which there were three different kinds), galmey, zinc, and magnesia or brownstone. We must not think of these substances as corresponding to our present-day notion of purified chemical elements, but rather as names that were attached to naturally occurring mineral specimens of little value, failed metals freaks, as it were.[34]

Thus, the concept of semi-metals proceeded from the theory that metals were generated, or composed, from simpler substances. Late seventeenth-century and early eighteenth-century theorists such as Georg Ernst Stahl and Hermann Boerhaave also assumed that metals were composites. Stahl famously claimed that metals were composed out of a calx and phlogiston, while Boerhave held a more traditional view: that metals were composed of mercury and sulfur and that a true decomposition of metal would result in the obtaining of a mercury.[35]

Seventeenth-century chymists, as well as many of their predecessors, were aware that certain metals had a remarkable chemical stability and could be recovered intact even when they had seemingly disappeared during the various processes of the chymical laboratory. For example, Robert Boyle's well-known *Sceptical Chymist* of 1661 contained long discussions of the recoverability of gold.[36] Boyle's work, as well as that of many other earlier chymists, foreshadowed the eighteenth century's eventual rejection of the notion of metals as composite substances. But neither Boyle nor anyone else pursued this line of research with a determination that even approximated the work of the mineralogical chemists of the Bureau of Mines. During the course of the 1730s to 1750s, chemists such as Brandt, Scheffer, and Cronstedt came to regard all theoretical inquiries into the composition of metals as useless speculation of no utilitarian value. As we will see, this position stemmed from their intimate familiarity with assaying and natural history, as well as from their cameralist inclinations. There was really no reason, stated the Bureau's chemists, to look for constituent parts of metals. As metal was the end product of the production process at mines, so it should also be regarded as the end product of chemical analysis. Furthermore, the isolation of a previously unknown metal or semi-metal was to be considered the discovery of new chemical entity, that is, a new species of metallic substance.

The articulation of the concept that metals were basic species proceeded from mechanical philosophy, which assumed that material reality was a machine composed of discreet, identifiable parts that could be combined and recombined. When applied to mineralogy, this philosophical position received a new empirical and economic foundation. It also introduced a new goal to chemical and mineralogical investigation: to find pure metals in samples of composite minerals, and to define and compare the properties of the new metals with the properties of those that were already known. In a sentence: the ultimate goal of *chemistry* was reduced to that of *assaying* for metals followed by natural historical description. From the point of view of curious—rather than utilitarian—investigators of nature, the position was quite radical. Pott, for example, pursued a similar research program of investigating the composition of minerals. Working in many ways parallel to the Bureau chemists, he conducted, among many other things, careful investigations of zinc and bismuth.[37] But Pott, a fervent adherent of the phlogiston theory who had come to chemistry from the field of medicine, did not allow that the search for metals should take precedence over the study of other mineral substances, such as clays or alkaline earths. He did not approve of the method of the *Bergjunge* (mountain boy) who "searches out that which is shining and metallic, but . . . throws away the matrix, or mixed-in species of stone."[38] The passage, published in 1751, may have been a criticism of Cronstedt's claim to the discovery of nickel, which was published the same year in *Transactions of the Swedish Royal Academy of Sciences*. It can also be interpreted as a critique of earlier works by Brandt.

It was Brandt who had taken the new approach into print in his 1735 paper on cobalt, "Dissertatio de Semi-Metallis" in the *Acta* of the Uppsala Society of Science. Brandt began his paper with a simple definition. A semi-metal should be defined, said Brandt, as a substance that has the general form, color, and weight of a metal but that was not malleable when beaten by a hammer.[39] He distinguished between, on the one hand, pure substances that could not be further decomposed, and all others. Metals and semi-metals (among the latter he counted mercury, bismuth, regulus of antimony, cobalt, arsenic, zinc, and cast iron or "ferrum crudum") were pure; all others were composite: "It is

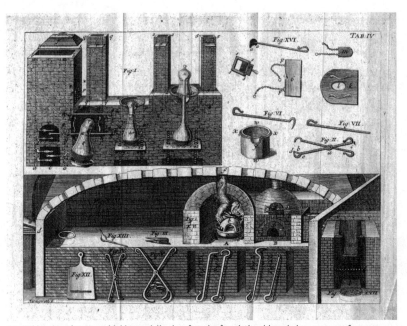

FIGURE 9. Assaying was a highly specialized craft and a foundational knowledge necessary for any mining operation. These furnaces and tools for assaying are illustrated in Johann Andreae Cramer, *Anfangsgrunde der Probierkunst* (Stockholm, 1746). Reproduction by Tommy Westergren, Library of the Royal Institute of Technology, Stockholm.

evident from the definition of semi-metal given above, that vitriols, cinnabar, minera, that is, metal bearing veins, as well as earths and glasses [vitra] of that kind in which nothing pure [*gediget*] can be found, cannot rightly be held as semi-metals or metals."[40]

Brandt's discussion of metals and semi-metals cannot be said to have been an innovation. First, it drew on age-old assaying practices. Consider, for example, the course in assaying held at the Bureau of Mines by Jacob Fischer, a student of the Freiberg chemist and mineralogist Henckel. Fischer's was a practical, straightforward, hands-on course. He began by presenting tools and materials, and also giving advice on how to make them. He then moved on to methods by which one could assay for specific metals in mineral samples. He also taught how to assay for substances other than the traditional metals.[41]

Second, Brandt drew on a long tradition within chymistry itself by using what has been called the negative-empirical concept of a chemical element.[42] According to this ideal, a substance was perceived as elementary up until the point that current analytical methods could decompose it. As William Newman has put it, "the concept is negative in that it defines an element solely in terms of what chymistry cannot do, and empirical in that it relies on the experience of the laboratory."[43]

Brandt's innovation was not methodological, nor did it derive from his use of the negative-empirical concept. It consisted in his novel way to combine these two for purposes of mineralogical taxonomy. He wanted to establish that cobalt belonged to the category of semi-metal. In the main part of his paper, Brandt discussed the differences between the semi-metals, in particular between cobalt and all others. By use of chemical trials, he established that cobalt could not be considered the same substance as any other semi-metal. He also presented a series of trials that clearly distinguished cobalt from bismuth.[44]

Brandt's mode of reasoning and methods were taken up by his colleagues. His paper was to become the first of a series that presented and defined new metals, issued from chymists at the Bureau. Of these, the most well known is Axel Fredrik Cronstedt's "Rön och försök gjorde med en malm-art, från Los kobolt grufvor i Färila Socken och Helsinge-land" (1751), in which he claimed the discovery of nickel.[45] In a similar vein, Henric Teophil Scheffer presented a paper on platinum, titled "Det hvita gullet, eller sjunde metallen. Kalladt i Spanien Platina del Pinto" (1752). It is considered the first thorough chemical investigation of the metal.[46] Others included Benct Qvist's paper of 1754, which claimed that a sample of black-lead (*blyerts*) that he had examined contained, in all likelihood, a previously unknown metallic entity. Qvist, however, did not follow through the investigation and failed to isolate the new metal. This was done a few decades later by Scheele (1778) and another Bureau chemist, Peter Jacob Hjelm, who was the first to obtain molybdenum in a metallic state (1781).[47]

These papers proceeded from the negative-empirical concept of an element, and used it in a well-conceived and theoretically articulate way to argue, first, that there was a set number of previously known metals and semi-metals, and second, that new discoveries had been made that added to the list. These were straightforward claims. As, for example, Cronstedt stated: there was no known alloy or composition with properties similar to that of nickel. For this reason nickel "should be regarded as a new semi-metal" until someone could present a method to compose it out of the twelve known whole metals and semi-metals.[48] Similar passages can be found in the works of other chemists at the Bureau. The group clearly shared similar opinions. They based their investigations on the works of one another and defended each other against critique.[49]

The grand old men of the group were Georg Brandt and Henric Teophil Scheffer. Both men were able teachers who transmitted their

views and methods to a comparably large group of students. Scheffer's lectures are well known, as they have been published, annotated,
and expanded by Torbern Bergman. From what remains of Brandt's
lectures, it is clear that he introduced his students to his latest theories
and discoveries. In addition, Brandt taught a comprehensive critique of
older theories of matter. He emphasized that investigators who claimed
to know the component parts of material objects needed to prove such
claims by following up their analysis with synthesis. Chymists knew, he
claimed, how to use fire to reduce plants and animals to salts, oils, spiri-
tum, water, and earth. From this it did not follow, however, that these
substances were the component parts of plants and animals. The reason
was that fire caused not only a separation of substances but also another
"unknown change, decomposition and mix-up . . . that is entirely different from the composition observed by nature in the making of plants
and trees." If a tree was to be reduced to its component parts, it must
first be reduced to prime matter. But that seemed impossible, or at least,
chemists were completely ignorant of how to achieve it.[50] The three
principles of sulfur, mercury, and salt received similar critical treatment.
Brandt's disdain is apparent. The speculation of "some Paracelsian[s]"
was described as "impossible and ridiculous" guesswork. Chemists
should acknowledge no principles that were not derived from proven
causes and proper experiments, and no one had ever, claimed Brandt,
been able to isolate sulfur or mercury from animals by way of experiments.[51] Brandt's influence on the following generation of Bureau chemists must not be underestimated. As an example, it can be mentioned
that Cronstedt's first lesson in Brandt's laboratory was to isolate cobalt
from a mineral ore.[52]

The claims of the Bureau of Mines chemists did not remain unchal-
lenged. Justi's dismissal of Wallerius has already been mentioned. He
was similarly critical of the Bureau chemists. Without mentioning any
names or texts, he attacked Brandt's investigations on how to distinguish bismuth from cobalt. As bismuth was a proper semi-metal and
cobalt was not, any such investigations were unnecessary. Cobalt, said
Justi, consisted of a mixture of iron, copper, bismuth, and arsenic.
Nickel too was of a similar composition. In a passage that must have
been particularly insulting to the pride of the Bureau chemists, Justi
claimed that cobalt ores had not yet been properly investigated. As he
put it: "A thorough study of the Cobalts . . . would be a great merit for
a meticulous chymist."[53] It goes without saying that no Bureau of Mines
officials were included in the group of meticulous chemists that Justi
invited to help him further mineralogical knowledge.

Justi's position was perceived as an attack on Swedish honor, and Brandt, Cronstedt, and others retaliated. Brandt responded in an oration held to the Swedish Royal Academy of Sciences, and Cronstedt penned an anonymous response, published in the periodical *Den Swenska Mercurius*.[54] Cronstedt also criticized Justi in a barbed passage in the foreword of his own mineralogy.[55] Patriotic emotions ran high, but in the end, Justi's attacks occasioned no reevaluation of the claims of the Bureau chemists among natural philosophers in the Holy Roman Empire or elsewhere.[56] The house journal of the Göttingen professors, the *Göttingische Anzeigen von gelehrten Sachen*, even defended the Bureau chemists' critique of Justi on the grounds that he had provoked the Swedes.[57] The status of cobalt and nickel as semi-metals was established quickly and with a minimum of controversy. But these discoveries were of only marginal importance by themselves. It was when they were integrated into systems of mineralogy that the full meaning of the new ontological category that they represented would become evident.

Order and Mineralogical Systematization

Minerals had long posed a peculiar challenge to natural history. Most mineral specimens were decomposable into constituent parts that were known to be other mineral specimens.[58] They were clearly not species in the same sense as botanical and zoological objects but rather aggregates or composite substances. But what were they composite substances *of*? As we have seen, the chymical principles were elusive entities and afforded little help to mineralogical taxonomy. Without an agreed-on concept of species, a mineral system based on composition was a very difficult enterprise. Pott, for example, did not think that chemists' knowledge of mineral composition was sufficiently advanced as yet to be used as the foundation of a mineral system.[59]

Linnaeus and many others sidestepped the problem by advocating a system analogous to that of botany and zoology. Minerals should be ordered according to their outward appearance, that is, grouped after color, structure, and so forth. The mineralogical chemists at the Bureau welcomed Linnaeus's general ambition to assign every natural object to a specific category in a theoretically all-encompassing system, and to provide each with an unambiguous description and name. But they found Linnaeus's emphasis on external characteristics problematic.[60] True to their background in assaying, their main interest in mineral systematics was as a help in identifying pure metals—the commercially

valuable content of minerals. A taxonomy based on external characteristics was not conductive to this aim.

The mineralogy (1747) of Uppsala professor of chemistry Wallerius can be characterized as a midcentury attempt to bridge the gap
between chemistry/assaying and mineralogical taxonomy. It combined
a thorough knowledge of chemical analysis with the Linnean aim of
achieving a total systematic overview of the mineral realm. It was an
influential work. Dubbed by Theodore Porter "the representative mineralogical treatise of the first half of the eighteenth century," it would
be translated into at least five languages.[61] There were, however, several
major differences between the mineralogy and chemistry of Wallerius
and that of the Bureau of Mines group. Wallerius was in many ways a
thoroughbred Linnean. He emphasized didactic overview and width,
as well as practical utility. Laboratory experiment and analysis held no
privileged position. From this perspective the Bureau of Mines chemists were only active in a small chemical subdiscipline: that of chemical analysis of minerals.[62] Cronstedt was a student of both Wallerius
and Brandt, and the person who would unite the Uppsala approach to
systematic natural history with the analytic tradition of the Bureau of
Mines.[63]

Cronstedt was completely opposed to systems based on external characteristics of minerals. In his view, his system was far from
complete, but it could function as "a bar or opposition to those . . .
who, *entirely taken up with the surface of things*, think that the *Mineral Kingdom may with the same facility be reduced into classes, gen-
era, and species,* as *animals* and *vegetables* are."[64] According to Cronstedt, the mineral bodies that were found in nature were almost always
composites or mixtures of a smaller number of pure substances. The
truth about mineral bodies—that is, all of interest that there was to
know about them—was the composition and proportion of pure substances that they contained. It came down to chemical analysis to reveal
this truth, and the outward appearance of the minerals gave no real
guidance.[65]

Cronstedt's system was completely integrated with his chemistry. To
use composition as a main tool to make distinctions among minerals, it
was necessary to have a clear idea about what constituted a pure chemical substance. This was provided by the negative-empirical concept of
an element, and by the list of pure metals and semi-metals proposed by
the Bureau chemists. His work demonstrated how chemical knowledge
could be used to define simple substances as species, which in turn could
be identified as some of the components that made up the more complex

minerals. But Cronstedt was acutely aware that his system was incomplete. For example, he desisted from pointing "out a particular earth for each kind of stone," claiming that chemical knowledge was not as far advanced as to permit it.[66] His system nevertheless indicated how a more complete system should be developed.

In this way, Cronstedt also pointed the way to what he considered worthwhile further research. The goal of chemistry was to discover new metals and other simple substances, and the goal of mineralogy to ascertain in what proportions minerals were composed from them. Both substance discoveries and mineralogy were part of an overarching project: to increase the revenues of the mining business. The system was economically useful precisely because this type of chemical analysis had its roots in the assaying of mining produce. The end product of analysis was salable goods, which also was the end product of the mining process.

On the other hand, the search for the component parts of metals was reduced to an idle waste of time. The final paragraph of his mineralogy challenged other chemists to refute his claim that his was the most proper and fruitful way of doing chemistry.

> There is no danger attending the increasing the number of the metals. Astrological influences are now in no repute among the learned, and we have already more metals than planets within our solar system. It would perhaps be more useful to discover more of these metals, than idly to lose our time in repeating the numberless experiments which have been made, in order to discover the constituent parts of the metals already known. In this persuasion, I have avoided to mention any hypotheses about the principles of the metals, the processes of mercurification, and other things of the like nature, with which, to tell the truth, I have never troubled myself.[67]

As we see in the next chapter, the open challenge posed of alchemy and astrology in this passage was no coincidence. When chemistry and mineralogy were reinterpreted as cameral sciences tightly connected to the sphere of useful production, the available space for curious speculation was narrowed down. It was not only alchemy that was cast away but also the entire quest for the principles of the metals. Neither had been proven false: they were considered futile because they were economically irrelevant. Here was a brave new research program indeed.

: : :

Four conclusions can be drawn from the creation of mineralogical chemistry as a part of the cameral sciences. The first concerns the claims about the special pure status of a specific set of metals and semi-metals, and their systematic ordering in relation to other substances. A thorough systematic and chemical knowledge of the constituent parts of matter would be created only if this program was expanded. By discussing metals as species in systems of natural historical knowledge, the chemical mineralogists of the Bureau could begin to establish a measure of epistemic hegemony in the mineral and chemical realms. Their claim—issuing from an essentially mechanists' ontology—was that only pure substances that could be conceived of as material components were worth searching for. The claim reaffirmed the importance of composition as a central research area for chemists and was the original formulation of a research program that would engage a great number of European chemists in the centuries that followed.

Second, the information-gathering, patriotism, and day-to-day useful pursuits of the Bureau's officials contributed to the turning of chemistry and mineralogy into tools in the service of the Swedish state, facilitating the exploitation of mineral riches and increased control over production. What we see from this point of view is an adjustment of the goals of chemistry and mineralogy to those of the state, as expressed in the Direktionsprinzip and through the cameralist system for governance. These fields of knowledge were transformed into aids for the development of the country's resources.

Third, I wish to emphasize the underlying economic rationale of this line of research. Metals were not simply lined up on the shelves of natural cabinets, but also on the shelves of merchants. The emerging concept of the chemical element allowed nature itself to become commodified. Nature, as symbolized through the stable commodities of cobalt, platinum, and nickel, was reduced to raw materials. When mechanical philosophy was fused with production, economic units were turned into epistemological boundary markers. They were allowed to delimit the borders of reality; chemical elements became an ontological category. With these elements in place, there was no need for chemists to consider, or even speak, of any entities or processes that did not have a market value.

It is important to remember that this was a limited, and in many senses limiting, conception of matter. From this follows the fourth aspect of this process. While the process outlined here essentially constituted the creation of a new, ontological category through which matter was understood, the Bureau chemists did not really contribute to the knowledge of the ultimate constituents of matter. From Odhelius, through Brandt and Cronstedt, the Bureau's mineralogical chemists abandoned the quest for the component parts of metals, and with it, speculation about the ultimate constituents of matter. As chemistry was evolving into a mechanical and useful science, the quest for objects that did not fit within the given constraints was deemed inappropriate. These constraints—that chemistry had to be mechanical and economically useful—were as much boundaries against untoward speculations as they were theoretical innovations. They also neatly tied mineralogical chemistry to what its practitioners identified as Enlightenment values.

6 Capturing the Laughing Gnome

If it is true what Locke did say,
That God did matter give
thought's perceptive ray
and therefore that no spirits live:
Then, his conclusion we must hear
That what we soul and spirit call
with death to sundering will fall
and with our lives will disappear.

HEDVIG CHARLOTTA NORDENFLYCHT, *Qvinligit tanke-spel* (1744)[1]

Would there be no more spirits, and did this mean that
our lives truly ended with our deaths? The poet Hedvig
Charlotta Nordenflycht shied away from the idea. So did
most eighteenth-century intellectuals and natural philosophers. But not everyone saw the problem and the ultimate
conclusion as clearly as John Locke and Mrs. Nordenflycht. If thought was the outcome of a process of matter
interacting with matter, it followed that soul and spirit
were mere material constructs, and thus neither of divine
origins nor eternal.

As this chapter shows, discussions of curious phenomena did not simply vanish from the Bureau with the eighteenth century. Instead they moved to a private sphere. The
public silence and professed disinterest that surrounded

theories of metallic transmutation was extended also to several other areas. No longer fit to be part of the hegemonic utilitarian discourse, premonitions and visions of angels, devils, and ghosts were now associated with religion, women, and family matters. Alchemy as well as magical practices and intercourse with the more physical keeper entities were decried as superstitions of the uneducated. Nevertheless, discussions continued in unpublished manuscripts, diaries, and private exchanges between close friends. This chapter discusses the boundary work performed by chemists and mining officials to seal off these subject matters from the sphere of productive and rational knowledge. It proceeded from a general association of superstitious beliefs and practices with women, the lower classes, those without formal education, foreigners, and ethnic minorities.

Significantly, Nordenflycht's poem, published in a volume titled *Female Thought-Play*, was a woman's transferal into the public of private discussions. She was no stranger to the sentiments of mechanical philosophy. In her youth she had been betrothed to the mechanicus Johan Tideman. He had introduced her to the recent philosophical literature and engaged her in learned discourse. Tideman could also let his imagination fly. Sometimes, she said, he "extended his powerful imagination to all the possibilities in nature, which could exist, [talking] about both spirits and bodies." Tideman's sweet talk with his as-yet-not-famous fiancée about spirits and material bodies was a mainstay of the intellectual environment to which he belonged. He was an Uppsala graduate and a student of Polhem.[2]

It is important to remember that not every form of mechanical thinking current in the seventeenth and early eighteenth century may be subsumed under the heading Cartesianism. There were many mechanical theories preceding that of Descartes, and even the great philosopher's adherents modified his system to suit their own needs. For many professed mechanists the theory was mainly used as a set of metaphors that limited discourse about nature to such phenomena as could be discussed using terminology from the world of crafts (grinding, cutting, mixing) and mechanics (ropes, pulleys, liquids flowing through pipes). The world of spirit, on the other hand, was not expressly excluded from their arguments.

Polhem framed hypotheses about ghosts and communication of emotions over long distances using a mechanical theoretical framework. We taste when salt dissolved in water touches the palate, and smell when breathing in salt dissolved in air. The vibrations of air give hearing, and the vibrations of the ether, sight. It is all a matter of corpuscles coming in contact with our sensory organs. In analogy, there could be a more subtle matter than ether that moves our thoughts in the brain. This would explain why children and their parents, good friends, and married couples sometimes sense each other over a distance of many miles. In a continuing analogy, he argued that ghosts also could be explained this way. They could be caused by the action of a fine matter on thought.[3]

Polhem's hypothesis was proposed in a manuscript, and it did not openly contradict Descartes. But it cannot be described as Cartesian: to suggest that the dead may walk the earth in shapes fashioned from subtle matter does take some of the edge out of Cartesianism. If the dead were around, albeit in subtle shapes that could only be detected by the sensory organ of the human brain, why not angels, devils, and perhaps other creatures? Who was to say that there was not a fully fledged spirit world for humans to interact with, all around us? This conclusion seems to follow from the proposition that the subtle shapes of ghosts were made from a matter that also functioned as a medium for telepathic communication between humans. This position would go against the spirit of Cartesianism indeed. Rather, Polhem's speculations resonate with the Neoplatonic conception of bodies of light, or subtle matter. Let us compare it to the influential late seventeenth-century Neoplatonism of Anne Conway. According to Conway, spirits were subtle and ethereal bodies. They could penetrate each other and were ultimately one substance. Therefore they were not limited in their actions by material extension.[4] To Conway this formed part of the explanation for the telepathic phenomenon that Polhem sought to explain: "if two Men intirely love one another, they are by this love so united, that no distance or place can divide or separate them; for they are present (with another) in Spirit; so that there passeth a continual Efflux, or Emanation of Spirits, from the one to the other, whereby they are bound together, and united as with Chains."[5]

The young Swedenborg, too, framed experiences of similar phenomena in mechanist metaphors that made use of a concept of subtle matter. Arguing for transmission of thought at a distance, he hypothesized that there were thought-vibrations that moved through space analogous with the movement of sound and light. This was because there were vibrating membranes in the brain, which could pick up the movement of other membranes that vibrated to the same pitch.[6]

As these examples show, self-proclaimed mechanical philosophers held on to the possibility of action at a distance through the medium of subtle matter. There was a limit to radicalism. After all, the mechanical philosophers' views were based not only on mechanical explanatory models and empirical rationalism, but also on a solid bedrock of Christianity, as well as on a long tradition of natural philosophical thought. As Copenhaver observes, the concept that "fine material substances or gross spiritual substances—*spiritus* or *pneumata*—can and must be used to explain coherence and continuity of action among natural phenomena" had an impeccable pedigree. It was part of an explanatory context for natural phenomena that had been developed by the most prominent authorities imaginable, such as Plato, Aristotle, Galen, Ptolemy, Plotinus, Avicenna, and Aquinas.[7] Given this background, it is not surprising that an openness toward subtle matters, and the phenomena that came with them, continued among many natural philosophers into the 1750s and beyond. Theories about vital spirits and subtle matters could blend seamlessly into mechanical philosophy but came at a price. They left open the theoretical possibility that a host of angels, devils, ghosts, and trolls inhabited the mechanical universe. And furthermore, that such entities were perceptible through the sense organ of the human brain. While the proper objects of physical and chemical investigation were purely material, entities composed of subtle matters could very well have a place in an essentially mechanical and material universe. As they were useless objects of knowledge, it would be idle to speculate about them. Nevertheless they would be useless objects that existed.

The first stirrings of radically dissenting views appeared toward the close of the 1750s. I have already quoted some of Cronstedt's pointed attacks on various aspects of mystical science from his 1758 manuscript. Dripping with irony, Cronstedt's text proceeded from a taxonomic systematization of the various branches of mystical philosophy, or "Philosophia Naturalis Mystica." He accepted three main branches: astrology, magic, and alchemy. Of these, magic and alchemy had further subdivisions.[8] Cronstedt's text is a remarkable exercise in boundary work. He went to great pains to relocate these beliefs to marginalized groups in society. These were children and youths, the ethnic minorities of Finns, Sami and Germans, the lower classes, and women. The mystical sciences were also associated, through the cabala, with Jews.[9] He began by mixing up the views of his opponents into an essentially incoherent sludge of weird views. Taking on the voice of Odin, the old Norse god of magic and war, he proclaimed that Sabbath witches really were alchemical wizards who held on to the belief in old Norse gods and used the philosophers' stone as a means of instant transport: "For, by the force of your sharp common sense, you can see that a substance which can move the body of a human in a moment (that is less than in the blink of an eye) from Åhl [in the province of] Dalarna to [the island of] Blåkulla off the coast of Öland, must in your worthy hands, have the ability to be used for transmutation, and for universal medicines."[10]

Cronstedt then pretended to regain his senses and proclaimed his whole system as madness and charlatanry. Nevertheless it is clear that it was a species of madness that Cronstedt had spent a long time studying. Not only had he had a few discussions with practitioners, but he had also read a great number of books on the subject, and had conducted a fair number of magical trials in his youth. Cronstedt's knowledge of the subject was both intimate and personal; shame and disappointment may perhaps partly explain the force of his rejection. As a child he had listened faithfully to the stories of magic told by his family's maidservants, and as a young man he had partaken in a number of magical trials. Hence, it was easy for him to associate magic with the undeveloped intellectual powers of children and youths. He had been present or close by, as he put it, when a frog had been put in a bored-through box in an anthill. After a while the skeleton had been removed, and two bones, one shaped as a hook, the other as a spade, had been preserved. These were then used to, respectively, attract and repel individuals "of the fair sex." In another trial he had collected the points of nine knives that had been accidentally broken. From these, he had intended to make a new knife, which he would have used to disclose the presence of evil witches. The plan failed, however, because he had been unable to convince a blacksmith to forge the magical blade for him.[11] Cronstedt associated interest in these matters not only with inexperienced youths but also with minorities and women. I now discuss these in turn.

The Irrational Others

Among Swedes—the majority population of Sweden—the use of magic was often associated with the country's Finnish and Sami minorities. Cronstedt explained his association of mystical science with the rural people's use of magic by proclaiming both to be parts of the same tradition, which originated from the Jews, only the magical knowledge of the Finns had been confused and degraded during the Babylonian captivity. It is clear, however, that Cronstedt had gained a more advanced knowledge of rural magic than that of his childhood trials, through conversations with Finns about the abilities of cunning. In this way he had learnt why the horses pulling his carriage had magically stood in place one afternoon in 1747. At the time, Cronstedt, still an auscultator, had been on a journey with his main patron at the Bureau, Daniel Tilas. Suddenly their horses had refused to move further on the forest road. The travelers returned to the farmstead where they had previously changed horses for new ones, but the second set also froze in the same spot as the first. Only the third set of horses managed to move beyond the fateful spot, and only at eleven o clock at night. Long after the event, two Finns had explained to Cronstedt that he had been stopped by an evil spirit in the shape of a bird. It had been set upon him by a cunning who wanted to keep the mining authorities out of the forest, for fear they would build new ironworks there.[12]

Experiences such as this one clearly made an impression on Cronstedt, even to the degree that he almost dropped his irony for a more objective narrative style. Nothing of the kind can be said of his denigration of alchemy. It too was associated with a minority: the Germans. While he described magic and astrology as common, Cronstedt described alchemy as belonging to the better classes of society. He stated that it should properly be seen as one of three separate branches of "mystical" (i.e., secret) science, where the others were astrology and magic. But although alchemy held on to a good reputation in finer society, he really suspected that "she" was a pimped-up daughter of her common mother, magic. Alchemy was rhetorically skilled and spoke several languages. She was, however, best versed in German. Just as Cronstedt described alchemy as a low-class immigrant woman pretending to be someone better than she was, the practitioners of her art were bunched together into a suspect lot of foreign extraction. The main parts of alchemy were alchymia rutomanica, or dowsing; alchymia medica, or the quest for the universal medicine; and alchymia metallurgica, or gold making.[13]

Through this systematic division, alchemy became a warped shadow of his vision for chemistry. As chemistry was the preeminent science for the discovery, extraction, investigation, and sale of metals, so irrational alchemy shadowed its rational sibling. Rutomanica, that is, dowsing, was concerned with the discovery of metals, just as gold making was concerned with their creation and sale. The association of dowsing with alchemy is unusual.[14] In Saxony, dowsing's reputation as a useful practice remained intact throughout the eighteenth century. The most renowned were the Freiberg dowsers, who had high status, and who did not, for example, perform their own digging. Indeed, the town was renowned for dowsers and unwilling to part with them to other mining areas.[15] In comparison, the Bureau of Mines made little use of dowsers, and there was no community or group of reputable dowsers in steady employ by the Bureau. Maybe Cronstedt held dowsing to be alien to the

Swedish setting and considered it a foreign, disreputable art. Through dowsing, alchemy was further associated with disreputable foreigners, and in particular, with German projectors.

In an effective anecdote Cronstedt combined the theme of the fraudulent German dowser with mocking the belief in mining spirits. There was a German dowser, he said, who developed a mine for a consortium of Swedish owners. In the presence of his employers, he had claimed to know that the mother lode was close. To convince them, he cried down the mine shaft: "Old Mountain Man how far is it now to the ore?" Upon hearing this, the mine keeper (or as Cronstedt intimated, the German's servant) yelled back that it lay at three hundred fathoms. The foreignness of the dowser was emphasized as his words were rendered in faulty Swedish, as if spoken with a pronounced German accent.[16] This highlighted both the man's status as an unlearned foreigner and the incomprehensibility of his activities.

The passage can actually be read as even more comical, and damning, than it appears. The character of the German market crier (*Marktschreier*) was something of an established figure in eighteenth-century Swedish comedy. The market crier appeared on stage showing pictures, or illusions, in a laterna magica. Greeting his audience with a characteristic outcry of "Ho ha!" he accompanied his show with a flow of German commentary mixed up with Swedish words. To a cultured and theatergoing reader such as could be found among the officials of the Bureau, Cronstedt's anecdote may have evoked the character of the German market crier and his laterna magica, effectively reducing the German's knowledge of mining to a comedy performed by a peddler of illusions.[17]

Cronstedt's pun was part of a remarkable turnaround in the perception of itinerant foreigners by the Bureau, which, as seen in the previous chapter, had been founded mainly by Germans. Although previously they had been a highly sought after resource, now German projectors and artisans in search of employment had come to represent outdated types of knowledge that no longer appealed to the Swedish mining authorities. Intensive contacts with high-ranking foreign mining officials, high standards of formal education, the many foreign study trips conducted by officials, and access to advanced academic and technical literature had now completely replaced immigration as a source of knowledge about how mining and smelting were conducted abroad.

The itinerant foreigners and their knowledge of traditional mining and smelting practices were transformed from a resource to something suspicious and problematic. There was also the aspect of social standing. The Swedish officials were often noblemen and held high status in society. The foreign travelers were low status in comparison, and their families, education, and previous experience were all unknown. There was no one to vouch for the trustworthiness of their knowledge.[18]

In 1747 Samuel Troilius, master of mine at Falun, wrote an answer to a petition to the Parliament, handed in by the German smelter Paul Fredrik Köppen. Köppen had proposed a new method for the improvement of the smelting at the mine. Troilius wondered why Köppen, if he was such a wonderful smelter, could not find employment in his own homeland. Then he mentioned the names of eleven previous foreign project-makers who had already failed to improve the smelting in Falun. He was of the opinion that most foreigners failed to realize that the Falun ore gave a much lower yield than that of most other copper mines. Therefore, the traditional process developed locally since medieval times was most likely the best one. Troilius also emphasized that if an unknown Swede showed up in Germany and offered to improve the smelting process in, for example, Mansfeld, he too would be seen as a charlatan.[19]

Mining officials like Troilius and Cronstedt could associate itinerant artisans with irrationality, fraudulent behavior, and untrustworthiness because they were secure in the knowledge of their own competence. Troilius's argument was highly informed, drawing on deep knowledge about mining matters, both foreign and domestic. In this milieu, there was only a thin line between foreign itinerant artisans and charlatans. Display of beliefs and practices that had begun to be associated with women, ethnic minorities, rurals, and the lower classes only served to make them further suspect.

Skepticism and the Rationalization of Work

If we stop at the remarks made by Cronstedt so far, we are left with an impression of him as a witty, ironical, and steadfast proponent of Enlightenment values. But it is not that simple. Cronstedt's letter was in all likelihood addressed to his friend and patron Tilas, at this time a mine councilor, the highest rank within the Bureau save the presidency. Tilas wrote a reply to Cronstedt in as witty and easy a style as that of his friend. But just like Cronstedt, Tilas also showed how enmeshed the practices they discussed were in their everyday pursuits in the mining business. In a delightful anecdote, Tilas told of how he, in his youth, had spent three full months walking and running through the wilderness, half starved and wearing ragged clothes, together with a dowser called

Holmberg. Tilas's task had been to inspect the dowser's art, take notes on interesting locations, and perform assaying tests.[20]

Cronstedt too told of events from his working life. In 1749 he had been ordered by the Bureau to clean up and inspect an abandoned silver mine, which had been found again and was to be reopened by an old sergeant named Blomberg. Blomberg also brought servants or workmen: two gardeners from Stockholm. As the group approached the village closest to the alleged mine site, Cronstedt began to suspect that his three charges would try to elude him. They lodged in a farmstead, where Cronstedt placed the three men in an inner room and slept himself in the outer one, where he had control of the inner room's door. He had been right in his suspicions, for during the night the men escaped through the window. Cronstedt decided, bravely and perhaps foolhardily, to follow them. In the middle of the second night of pursuit, he came upon them at an ancient cairn named the Robber's Hole. The men refused to answer his questions but silently gave him garlic and castor, both strongly smelling substances that presumably served as magical defenses. In the morning they explained to him that they had lain there hoping to spy the keeper of the mountain, who would tell them where the mine was. Blomberg said that he had met her before and that she was a beautiful woman, but the last time he had seen her she had taken on the hame of a bear. Cronstedt then asked if he spoke metaphorically, as when one calls someone who hangs around in kitchens looking for extra servings a kitchen bear. On hearing this the sergeant became angry and exclaimed that such free thinking ought to be punished. Cronstedt concluded the anecdote by saying that as his instructions from the Bureau had been to inspect a mine that had already been located, he left the group in the forest.[21]

As indicated by the anecdote, keepers of mines and the magical practices used to communicate with them were strongly pushed out of the sphere of useful knowledge, the same sphere that had been claimed by mechanical philosophy and later by mechanical chemistry. This ridicule was as much a part of the rationalization of work as the systematic overview, improvement, and control of mining processes. The new breed of mining officials at the Bureau had little use for interactions with keeper entities. They were, however, ridiculing a system of beliefs with old and deep roots, and one that was widely acknowledged as valid among their contemporaries. Pacts with keeper entities seem to have been a predominantly male strategy associated with male work, in particular fishing, hunting, and mining. In exchange for respect, gifts, and services—sex, food, drink, body parts, or one's soul—these pacts gave access to game,

FIGURE 10. Waterwheel and men at work at Älvkarleby fors in 1768. From Daniel Tilas's collection of drawings and prints. Note the sad-looking naked man in the foreground. His trident is a classical attribute identifying him as a creature of the water. He is probably a representation of the keeper of the stream (*Näcken*). Reproduction: Jens Östman, National Library of Sweden. Call number KoB Tilas II: 317.

fish, or ore.[22] For many early modern miners, interactions with keeper entities were an important part of the production process. Myths about the first disclosure of important mining sites often included stories about how animals, for example, bulls or rams, scraped the ground with their horns or front legs to lay open the glint of metallic ore.[23] Such animals may have been perceived as no less magical than dragons or trolls. Just like Sergeant Blomberg's bear, they were physical manifestations of powers who dwelled in the wilds, and who communicated to worthy humans that a deposit of minerals was now ready to be taken for use by humankind. Interpretations, however, were rarely stable and fixed. The Falun copper mine was traditionally held to have been coincidentally discovered by a domestic animal, a he-goat by the name of Kåre. But simultaneously, the Lady of the Mine sometimes chose to manifest herself in the form of a goat or kid.[24]

These entities were forcefully pushed out of their environments through the retelling of stories, either in an ironic way or simply by providing rational explanations of events. Stories about bulls and goats

could easily be interpreted as founding myths or anecdotes of how ordinary grazing cattle revealed veins by coincidence.[25] Dragons could be
rejected out of hand, although they were still part of the belief system
of the common people. In a letter from Bureau official S. G. Hermelin
to Torbern Bergman, Hermelin told of how the villagers of Rättvik believed that shining lights seen on the mountain were made by dragons
and could signal the place of a mineral deposit. Hermelin was of a different opinion and explained the fire as caused by chemical agents.[26] A
functionalist explanation would have it that these beings served little
or no purpose for the scientific mining and surveying advocated by
the Bureau's officials, but the ridiculing of belief in these beings also
served to define how men should act and what they should believe.
It was part of the redefining of the officials' identities, and of mining
as the prerogative of rational men engaged in useful production. Mining was male knowledge associated with a strong and hearty constitution, and just like manly science it could be juxtaposed with female
superstition.[27]

There is a clear continuity from Hiärne in this regard. In chapter 2, I
discussed how Hiärne left the question of the existence of keeper entities
open, while simultaneously claiming that they had no influence over the
mining business. At the same time, he upheld a strong belief that invisible but benign powers kept an eye on his family matters. The attitude of
Hiärne's grandson Tilas was different only in his degree of skepticism.
While he cracked jokes about keeper entities and dowsers with Cronstedt, he held back on some of his beliefs concerning the influence of unseen powers.[28] When Tilas's reply to Cronstedt is compared with entries
in his diary and autobiography, it becomes clear that he believed he had
experienced a premonition of the death of Queen Ulrica Eleonora and
also of a beloved daughter. Furthermore, in 1747 he had learned about
the birth of his first son, while traveling far from home, by feeling his
wife's birth pains in his own back and loins at exactly the time she was
giving birth. The pain was so intense, he wrote, that his servant had to
massage him with a salve in his carriage.[29] Interestingly, Tilas believed
that he had inherited these unusual abilities from his mother, who was
a daughter of Hiärne. When Cronstedt's letter is read in light of Tilas's
response, it appears that Cronstedt's witty tone did not carry throughout the letter. In particular his experience with the bewitched horses,
which he had shared with Tilas, was described without irony, and in the
same style as any other report about factual events. This is perhaps not
so surprising. The experience seemed to defy rational explanation, and
would have been both eerie and chilling to anyone.

Tilas's and Cronstedt's accounts point to their difficulties in effecting a wholesale rejection of these culturally widely acknowledged, and partly self-experienced, phenomena. Their ready association of these phenomena with women gives a clue as to how the problem could be handled. Cronstedt's depiction of alchemy as a daughter of the low-born mother magic is an instance of the association of women with the disreputable other. Yet Tilas's association of premonitions, emotional telepathy, and ghosts with the women of his family show how the gendering of certain phenomena allowed for a degree of continuance in beliefs.[30] The phenomena were primarily disassociated from the productive sphere of manly work. The most ridiculous persons of all were men who, like the unnamed German dowser and Sergeant Blomberg, sought, or pretended, to communicate with mine keepers in order to receive guidance in economic endeavors. These practices were depicted as standing in stark contrast to the endeavors of the scientific mining officials, as described in the previous chapter. But while certain sets of practices and beliefs were disassociated from the sphere of work, others continued to linger around objects and relationships that were not directly affected by the discourse on utility.

Dealing with the Alchemical Heritage, ca. 1730–60

Brandt and other Bureau chemists had rejected transmutative chymistry in the 1730s, and from about that time it was considered a thing of the past. But still theirs was a soft rejection. No or few public denouncements were made, and a general tendency to emphasize continuities remained. There were a number of reasons why the chemists at the Bureau held back from public criticism of alchemy. The first is that the very successful do not need to bother much with criticism of the less so. The Bureau chemists were too preoccupied with their own research program, and with discussions and debates about mineral analysis and taxonomy, to bother with what the alchemists were up to. There were other reasons as well. Let us again return to Cronstedt's views of alchemists in his 1758 manuscript.

> Our alchymists . . . how should they be known? Answer: . . .
> one should not search for them among such, who have learnt the
> operations and basics with mechanical chymists. Mineralogy is
> even less needed: For as I have heard from [the alchemist] Baron
> Hendrich Wrede, all of the printed works of Wallerius were of
> no use [to him] except some remarks, about the solidification

of water into earth or rock and on the mercurification of metals. Therefore when You see nothing of the chymistry of these times. . . . You are in the right company.[31]

At first glance Cronstedt's words show how far the separation between mechanical chemistry and alchemy had gone by 1758. A second glance reveals also the complex social context in which Cronstedt's statement was written. Regardless of their views about alchemy, Swedish mechanical chemists had to relate to practicing alchemists. Some of them, like Baron Wrede, a county governor and prominent politician, were influential men. Cronstedt, as he was well aware, needed to be careful in his criticism, unless he wanted to run the risk of upsetting important people. In his text Cronstedt also listed the names of six alchemists "with honest intentions" whom he knew about. These men were from the nobility: one was a general, one a colonel, and three were highly placed officials in the civilian state apparatus. He also mentioned some other men who were said to have believed in alchemy, worked on it, but had given it up. These too were prominent pillars of society.[32]

Among the names, one stands out—that of Gustav Bonde (1682–1764). Count Bonde was one of Sweden's most influential politicians. In the 1730s he had been the second most influential man in the country, and he was the twentieth generation of his family to hold the highest position that a Swedish nobleman could aspire to, that of councilor of the Realm. He was a part of the innermost circle of the government until 1738, and a well-respected figure even among his political opponents after his and his party's fall from power that year.[33] Bonde led a public life in service of his country. Among many other offices, he had been president of the Bureau of Mines between 1721 and 1727, the period when the Laboratorium Chymicum was reconstructed by Brandt. Bonde's support is apparent from an entry in the politician Nicodemus Tessin the younger's personal notes. It records an informal approach from Bonde, who had argued for the utility of a chemical laboratory attached to the Bureau.[34]

Furthermore, as chancellor of Uppsala University in the 1730s, Bonde had proposed that the university should establish a chemical laboratory, although this did not happen until 1750.[35] The reconstruction of the Bureau of Mines laboratory, as well as the establishment of a chair of chemistry at Uppsala, thus had an enthusiastic supporter in the highest reaches of government. For mechanical chemists there was only that slight problem: Bonde was a devotee of transmutative chymistry. He also had a small circle of like-minded friends, consisting of the national *censor librorum* (censor of books) Niklas von Oelreich, as well as three or four other politically influential men.[36]

There were connections between Bonde's interests and the Bureau. For example, he supported attempts at the Bureau to convert iron to steel without loss of weight (a form of transmutation).[37] Furthermore, his relative and confidant Fredrik Lorentz Bonde became an auscultator at the Bureau in 1735. In the end, however, it does not seem that Bonde sought to implement a transmutative agenda at the Bureau. Fredrik Lorenz never advanced beyond the lowest position in its hierarchy but instead he became a chamberlain at court in 1745. Bonde himself, although he was president of the Bureau, was politically appointed as such by the Parliament and the Council of the Realm, and thus had not advanced through the ranks of the Bureau. A letter from Gustav Bonde to Fredrik Lorenz Bonde of 1752 indicates why Bonde did not act more forcefully as a patron of transmutative chymistry at the Bureau: he had never been particularly interested in "the mineralogical work, which demands unending adventurous work and well-trained skills."[38] Edenborg, author of a biography of Bonde, has also shown that he did not primarily practice traditional techniques of metallic transmutation but preferred a method in which the body of the operator was used as the chymical furnace. Apart from the two Bondes, another name on Cronstedt's list also had a position at the Bureau: Thomas Blixenstierna. Blixenstierna became an auscultator in 1733 and traveled in the mining districts of the Holy Roman Empire in 1738–45. He received the Bureau's grant for chemical studies in 1747 and became an assessor in 1749.[39] According to Cronstedt, Blixenstierna abandoned alchemy before he died in 1753, but he does not indicate when, or why, this happened.

Given the interests of these influential men, it is not surprising that Brandt, Cronstedt, and others who held intermediate positions at the Bureau did not engage in public debate concerning the futility of alchemy. It would have been highly inopportune to do so. As we have seen, however, there was a degree of critique in manuscripts, private correspondence, and lectures held for small groups of students. In the case of Johan Gottschalk Wallerius, professor of chemistry in Uppsala, it was somewhat different. Wallerius had a positive view of his discipline's past and was not afraid of saying so. In a letter to Torbern Bergman, Wallerius criticized Bergman's dismissal of alchemy, saying that "even if alchemical trials have failed for many who are less knowledgeable about the properties of metals, nevertheless, chemistry has mostly to thank alchemy for important discoveries. Maybe, for that reason, [alchemical trials] should not rightfully and without exception, be called *fancies.*"[40]

Wallerius's position was much more cautious than Bergman's, but it was by no means an alchemical one, as defined by eighteenth-century mechanical chemists. Indeed, his opinion resembles that expressed by Boerhaave in *Elementa Chemiae* (1732).[41] In spite of this, historians of science Sten Lindroth and Tore Frängsmyr have gone to some length to portray Wallerius as a scientific misfit and extremely old-fashioned.[42] Frängsmyr has even presented the view that Wallerius was an alchemist, by which he meant that Wallerius was a deviant in his own scientific culture and time. This latter view, however, is not supported by any eighteenth-century documents, or by any other scholar who has studied Wallerius's chemistry. To the contrary, Wallerius was numbered among the most internationally well-respected chemists and mineralogists throughout the eighteenth-century. If nothing else, his position shows that appreciation of chemistry's alchemical heritage remained a viable position among chemical practitioners all the way to the end of the eighteenth century.[43]

Swedenborg's Transgression

As the son of a bishop, Emanuel Swedenborg appreciated the materialist implications of mechanical philosophy. Having been a controversial appointee to the Bureau, he had established himself as a respected and useful man during the succeeding decades. He had participated actively in the reshaping of the Bureau into a prominent environment for knowledge production. Already in 1736, however, Swedenborg had begun a work on organization of the human soul, which he chose to publish anonymously in 1740–41. Biographers have taken this work as a starting point for his changing interests. From then on he would gradually lose interest in the material world and instead focus on theology and the world of spirit.[44] Then, in 1744, an event occurred that would change everything for him. Jesus Christ appeared and told Swedenborg that he had been chosen to renew Christianity. Three years later, in 1747, Swedenborg asked to be relieved of his duties at the Bureau and began in earnest his second life as an anonymous author of theological and spiritual works. In him, radical proponents of Enlightenment found a particularly hard nut to crack. Of what use was it to localize unwanted beliefs to women, inexperienced youths, immigrants, and forest-dwelling rurals when this man of good social standing, and with an impressive publishing record, suddenly claimed that he had full—and everyday—access to the world of spirits? To these four categories needed to be added an additional one: madmen.

Swedenborg's message was also theologically problematic. The Swedish church defended its turf aggressively. Heresy carried the death penalty, and religious deviancy was tightly monitored. In 1726, a law had been passed that outlawed all private religious gatherings. The conceived threat was radical Pietism, and the church struck down its proponents when it could. Pietists could be imprisoned for life or banished from the country. Even touching lightly on the sensitive issue of the authority of the Bible could get one into trouble. In 1746, the historian Olof von Dalin suggested that if the figures calculating the receding water levels along the country's shorelines were correct, Sweden might have been under water at the time of the birth of Christ.[45] This earned him a protest from the ecclesiastical estate, delivered at a session of Parliament in 1747. The subsiding of the Flood was the only permissible explanation for the "diminution of the waters."[46]

Swedenborg's old acquaintance Gustav Bonde held a more informed view of Swedenborg as a theologian than most did. Bonde owned a number of Swedenborg's theological works and knew he was their author well before Swedenborg stepped out of anonymity in 1760.[47] After reading them, Bonde concluded that Swedenborg's main innovation was the claim that all angels and devils had been the souls of deceased humans who had been transformed into an exalted state through their piousness, or had brought themselves to damnation through their evils. From this it followed that humankind was a kind of "colony or seminar" that fulfilled the function of providing souls to populate both heaven and hell. Bonde also found it remarkable that Swedenborg claimed to have seen, listened to, and learnt from the souls of the deceased, as well as angels and spirits of this world, and from other planets and stars.[48] Bonde's objections to Swedenborg's teachings were formulated from a Lutheran point of view. He pointed out that if one's spiritual state after death was a direct consequence of one's character and actions in life, then there could be no room for intervention by the saving grace of God. Furthermore, the thesis that all angels and devils had been the souls of deceased humans contradicted the Bible, which stated that the Serpent of Paradise had been created before humankind. Bonde also criticized the way Swedenborg interpreted the Bible.[49]

Swedenborg, however, had protection from persecution. He had several close relatives among Sweden's bishops, as well as connections at court and in the Council of the Realm. Bonde's friend Niklas von Oelreich, the national censor of books, was also an acquaintance of

Swedenborg's. Thus, he had an exceptional array of powerful friends. The situation was curious: persecution of Swedenborg's followers would eventually be mounted by the religious authorities, but Swedenborg himself was largely let alone. Without doubt, Swedish Lutheran orthodoxy, religious persecution, and Swedenborg's exceptional social position put something of a damper on religious speculation and debate. It is also important that Swedenborg published his works anonymously in Holland and later in England. This meant that such discussions of Swedenborg and his works as did take place in Sweden tended to be initiated in other European countries.[50] Consequently, public understanding, then as now, of Swedenborg's theology and experiences of the spirit world was not particularly nuanced. Instead, gossip about his extraordinary abilities and personal credibility fed into ongoing discussions about the relationship between matter and spirit.[51] Could Swedenborg really see and talk with spirits? Could he perceive events happening in other parts of the country? Could he find hidden treasures? These were the issues that interested the public, not the intricacies of his teaching.

These matters became an issue when Swedenborg's identity was revealed to the residents of his hometown, Stockholm. Early in 1760, rumors began to circulate that there was a resident of the city who could freely communicate with the spirits of the deceased and with angels. This man also claimed that the spirits of the dead continued their existence in the otherworld much as they had on earth. Among the things they did was remarry with other spirits. Soon the full story was out: it was Swedenborg. Tilas communicated the news to Cronstedt in a letter:

> It is Swedenborg, who has intercourse with the dead whenever he chooses, and who can inquire after his former departed friends when it pleases him, whether they are in heaven or hell, or whether they hover about in a third, nondescript place. . . . Queen Ulrica Eleonora is doing well; she is now married to another noble gentleman, and lives in a state of happiness. I am all in a flutter before conversing with him and hearing whom my late wife, Hedwig Reuterholm has married; I should not like it, forsooth, if she has become a sultaness.[52]

Tilas then made a visit to Swedenborg. He knew him, of course, as a former colleague and acquaintance. He returned with a modified but still highly skeptical attitude and reported to Cronstedt that he did not know whether Swedenborg was crazy or not. He would suspend judgment for the time being.[53] Soon there was a line of visitors and carriages outside Swedenborg's house, and most visitors tended to return with a similar impression. Swedenborg's conversations with spirits was thrilling news for the Stockholmers, and even to members of his outer circle of acquaintances, such as Tilas.[54]

What would the Swedish community of natural philosophers do with the prophet who had stood up to preach in their midst? Their gut instinct seems to have been to conceal and suppress. In a letter to Cronstedt, Tilas expressed his concerns that the news would spread to foreigners. "This information must not be spread abroad. . . . These things are known to thousands here; but I do not think it advisable that they should become generally known."[55] Secrecy was of course impossible. It remained for Tilas, Cronstedt, and their friends to tackle the issue in public. Swedenborg's coming out is probably what triggered a wave of public boundary work in the scientific circles that he had frequented. The boundary work of Swedenborg's former associates had two aspects. First, there was the social disassociation from suspicious individuals, activities, and teachings. Second, there was an identification of the natural world with the world of solid matter. As we have seen, these sentiments had been brewing quietly for a while. Although Sweden had had a parliamentary system since 1719, there was strict censorship of printed material until about this time. Political debate, too, had been mostly limited to handwritten texts circulated primarily in Stockholm, especially when Parliament was in session. The debate on superstition that now flared up among natural philosophers coincided with a radicalization of political debate and was made possible by relaxation of censorship in the 1760s, which culminated in the Freedom of the Press Act of 1766.[56]

Denouncing Alchemy

The gatherings of the Swedish Royal Academy of Science became an important venue. Already in September 1760, Cronstedt gave a speech at the Academy to commemorate the deceased Henric Teophil Scheffer. Scheffer had been the head of the Bureau of Mines chamber of assaying. Deviating significantly from the standard formula of such a speech, Cronstedt boldly proclaimed that in his life of Scheffer, "travels, promotions, and other plays or gifts of fortune, will to one part be missing, to one part take the smallest space."[57]

Instead the speech became an attack on alchemy in which Scheffer was used as a mouthpiece for Cronstedt's now radicalized views. Cronstedt attacked alchemy, and signaled that he was uninterested in discussing chemistry's mystical past. It had not had the foundation of a proper theory of nature, and therefore it was full of guesswork, incomprehensibilities, and dreams.[58] He also emphasized chemistry's connection with experimental physics. He carefully pointed out that Scheffer had had a thorough grounding in mathematics and physics before he started chemistry. Cronstedt went further than just trying to save Scheffer from allegations of connections with alchemy. His privately voiced supposition that Brandt might have done alchemical experiments was now gone. Even Hiärne was described as a man who wanted to use chemistry to provide support for Cartesian natural philosophy. Cronstedt surely knew that this was stretching the truth, if not an outright lie.[59] Finally, he widened the attack to a number of practices and phenomena. These were also associated with the religious superstitions of the Catholic Church: "Dowsing rods, amulets, the philosophers' stone and potable gold are the fabrications of [Catholic] Monks, without doubt to mimic similar things, that the pagans bragged about."[60]

Cronstedt disassociated Scheffer from religious deviancy, spiritist speculation, and alchemy. Scheffer, according to Cronstedt, did not believe that he could examine spirit entities in his laboratory, he never mixed metaphysics with physics, and he was an upright Christian who was free from mystical whims.[61] Whether this was true or not, the attempts of Cronstedt and others to rewrite history were successful, and others followed suit. Torbern Bergman, in his oration to the memory of the deceased Georg Brandt in 1769, noted that Brandt seems to have had an interest in transmutative chymistry. This did not lessen Bergman's scorn of such pursuits. Instead he made a frontal assault on transmutative chymistry, placing it in a context of the mad and futile: "Every science has its secrets, or madnesses: Geometry has the squaring of the circle; mechanics has the Perpetuum mobile; Astronomy [has] Astrological precognitions; Medicine [has] the Universal medicaments; Ethics [has] Platonic love . . . etc. Chemistry has several: to make Gold, precious stones, malleable glass, universal solvent, etc."[62] In his oration to the memory of Anton von Swab, given the same year, Bergman called alchemical theorizing "fancies, that usually have the bad luck to, when closely examined, turn into unfounded figments of the imagination."[63]

What then about Cronstedt's previous concerns that it was necessary to give alchemy "a room of its own so as not to anger anyone"? He had written the comment on May 20, 1758. On August 20 of the same year, Baron Wrede, alchemist, politician, and country governor had died on his country estate.[64] Bonde's influence was waning; he had long since retired from politics. Oelreich remained in office until 1767, but his influence was partially circumscribed by the appointment of Erik von Stockenström as chief justice (1758). Stockenström was a student of, among others, Georg Brandt, who had used the Bureau of Mines as a steppingstone for a political career. Considered one of the major shakers and movers in the influential Hat Party, he managed to become included in the inner circle of the Swedish high nobility. During his period as chief justice he oversaw the dismantling of state censorship and implementation of the liberal Freedom of the Press Act of 1766.[65]

Times were changing. As the older generation was leaving the scene of politics and patronage, it became safe and fashionable to slash connections between chemistry and alchemy, and to publicly decry all forms of superstitions.[66] The subject matter, however, could still be delicate. In 1766 Tilas took on the task of commemorating the recently deceased Bonde. In his oration, Tilas first talked about Bonde's youth and his political achievements. He then proposed to his listeners that they join him on an imaginary visit to Bonde, to see what he did in his spare time. He switched style and painted the picture of inviting his listeners into Bonde's home: "But where does our Count go now? We see him hurry into his Cabinet, to throw off his scarlet robes [of state]. . . . Let us follow! We find him at work in his Laboratories, preparing assaying furnaces, muffles and crucibles . . . preparing all kinds of mineral samples to find out their contents."[67]

After this visit to Bonde's chamber of assaying, Tilas followed the count into an inner chamber, his chemical laboratory. He continued, "and I can also reasonably understand, what the meaning of the inner chamber is, where [he] works with various other kinds of furnaces, retorts and recipients, conducting chemical investigations."[68] Behind Bonde's chemical laboratory, there was another door, and now Tilas claimed that he did not understand anything anymore: "furthest in I see another room, intended for deeper reflections, that I in my ignorance, do not dare to describe, and now I notice, although too late, my own lack of ability to explain to you, Gentlemen, . . . the deep insight into the hidden secrets of nature possessed by our Count."[69]

Tilas then sneaks into the inner room, and from behind the count's chair he steals a look at the papers on his table and sees a printed treatise: *Clavicula hermeticae scientiae* (Lesser key of hermetic science), from 1732.[70]

In this way, Tilas could both praise Bonde for his achievements as a public man and reveal his secret life. The device he used was a stroke of inspiration. Scarlet was the ceremonial color signifying a councilor of the Realm. When Bonde, in Tilas's account, threw off his robes, he left public life behind and entered another sphere that was, implicitly, disconnected from his high office. By talking about the three rooms of Bonde's laboratory, Tilas was able to distance Bonde the public man from Bonde the secret alchemist. The first room, the chamber of assaying, had its door open to the outside world, and the world of public utility. It was the closest room to Bonde's life as one of the country's leaders. It was dedicated to utilitarian purposes and connected Bonde with his time as a president of the Bureau of Mines. The middle chamber, or chemical laboratory, was partly dedicated to utilitarian purposes, and partly to the lofty realms of chemical theory and speculation. The third, hidden room was pictured as Bonde's secluded refuge, and as such its presence did not really harm anyone, did it?

Bonde, however, had never hidden his interests quite as carefully as Tilas implied, and his alchemy had intersected with his public career at several points in his life. Tilas needed to reveal Bonde as a closet alchemist precisely because his interests were not secret. Even in Tilas's narrative, Bonde's innermost secret was a *published* treatise (albeit printed anonymously). By revealing Bonde, Tilas's clever oration did not do much for Bonde's reputation for posterity. His later biographers have described him as cloven in two: on the one hand, the rational politician, and on the other hand, a scientific dilettante and weirdo. A recent researcher has even described him as monstrous.[71]

A few other fellows of the Academy influential enough or well enough known to have their life stories reconstructed posthumously were J. G. Wallerius, Linnaeus, and Swedenborg. Along with Bonde they have been described, by older historians of science, as partially or wholly irrational thinkers and depicted as exceptions in an otherwise rational era.[72] This chapter has shown the deep faults of such propositions. The religious and epistemological positions of these men were representative of mainstream views in their local community during the eighteenth century. As still rather *curious* natural philosophers, they carried on a continuing discussion about a host of strange empirically observed phenomena and strongly held beliefs that could not be discussed in purely materialistic terms. Several scholars have pointed out the similarities of Swedenborg's religious worldview with that of Linnaeus.[73] Similarities can also be found among Bonde, Wallerius, Tilas, and many others not discussed here. A common cultural point of reference was Pietist literature, such as Christian Scriver's *Seelenschatz* (Treasure of the soul). This immensely popular book gave voice to an intense belief in a spiritual world of angels that was deeply involved in human affairs. According to Swedenborg's father, Jesper Svedberg, Scriver's book was indispensable, the most important book after the Bible. The book was also read in the home of the young Carl Linnaeus.[74]

This was not just not a matter of religious and intellectual curiosity but also of practice. Polhem's and Tilas's telepathy and premonitions, and Cronstedt's magical trials blur our present-day distinctions between elite and popular culture. Even Swedenborg's ability to communicate with spirits was not considered as unique by his contemporaries as it was by his latter-day followers and detractors. There was, for example, Erik Tollstadius (1693–1759), a Stockholm vicar and central figure in Swedish Pietism. Already in his lifetime there were many rumors about his miraculous abilities. His sermons were to have been so powerful that they converted those who had come to the church intent on reporting him for heresy. He was to have disarmed a murderer with a look and had revealed hidden crimes through his uncanny foresight. Angels, appearing in the guise of small children, had saved him on several occasions when he had been in mortal danger. There were also contemporaries who claimed to see that angels always surrounded him.[75] Tollstadius's example serves to remind us that the concept of holy men in communion with angels was far from alien to eighteenth-century Protestants.[76] Natural philosophers discussing these and similar phenomena sought to formulate responses to the challenge of mechanical philosophy: that there was a rift between the world of spirit and the world of matter. In their attempts to make their world whole again, they were representative of their time and their culture, not exceptions.

The largely successful attempts of men like Tilas, Cronstedt, and Bergman to make natural philosophers such as Bonde and Swedenborg appear as weird exceptions contributed to making Enlightened discourse about superstition self-referential. Knowledge about trolls, ghosts, angels, devils, magic, and alchemy stemming from ethnic minorities, foreigners, women, rurals, artisans, and the lower classes had already been classed as unreliable. Now some of the knowledge produced by educated, well-published male natural philosopher was judged as unreliable too. There was no need to listen to *anyone* who claimed to interact with invisible or nonmaterial entities, or who pursued alchemy.

There was no one who was more successful in this recategorization than Immanuel Kant. Kant's encounter with the works of Swedenborg seems to have been a defining moment for him. What was remarkable about Swedenborg, from Kant's point of view, was his status as a credible and reliable witness, at least initially. Swedenborg had all the proper credentials of a trustworthy person. He was a hard-working male of good social standing, a confirmed patriot, and a renowned author of

useful scientific works. But Kant's initial curiosity would soon be exchanged for deep skepticism and satire. His book *Die Träume eines
Geistersehers erläutert durch Träume der Metaphysik* (1766) was a frontal attack on Swedenborg, presenting him as a madman.[77] For Kant,
even if Swedenborg used to be a trustworthy person, the nature of his
claims singled him out as mad. Kant committed, as Steven Shapin calls
it, "the ultimate incivility," namely, "the public withdrawal of trust in
another's access to the world and in another's moral commitment to
speaking the truth about it."[78] Older research has assumed that Kant
gave Swedenborg such a bad reputation that he put an immediate end
to further German philosophical interest in Swedenborg. Neither, however, had the last word. Although the Kantian position would be the one
that entered the philosophical mainstream, Swedenborg would have a
deep influence on the theology of the period.[79] Nevertheless Kant signaled, in his way, the end of the curiosity paradigm among elite natural
philosophers, and the beginning of the Nordic and German Radical
Enlightenment.

: : :

In traditional accounts of the Enlightenment, eighteenth-century
changes in perception of the relationship between matter and spirit have
been associated with growth of skeptical attitudes in larger cities. The
process has been described as linear and unproblematic. Traditional,
local attitudes and beliefs were challenged and in many instances replaced by more cosmopolitan and materialistic attitudes through the
growth of a discussion, in public and privately, about curious phenomena. This chapter has pointed to a much more complex development;
it shows a picture of circulation between different local contexts. It is a
picture similar to that drawn by Franco Venturi in his studies of Italy
and the Enlightenment. In a discussion of vampires, Venturi shows that
although the practice of exhuming and destroying the bodies of alleged
vampires was an issue for courts and lawmakers in Habsburg Hungary
but not Italy, discussions on why these practices should be curbed drew
on arguments from Italian debates on witchcraft. Both debates, in turn,
fed the radicalism brewing in midcentury Paris.[80] Similarly, the discussions by Scandinavian mining officials of witches, trolls, spirits, dowsing, alchemy, and the limits of matter drew on philosophical works
and public debates that took place elsewhere in Europe, but they addressed local concerns of whose knowledge one should trust, whom
one should employ, and how one should go about when conducting

one's work. Parts of these discussion recirculated readily to other contexts, such as the Swedish Royal Academy of Sciences. But an absence of public debate during much of the eighteenth century made expressions of Enlightenment sentiments in the Swedish realm highly local and largely unconnected. Enlightened beliefs were not only localized to specific individuals, groups, and contexts, but also to specific areas of knowledge, as held by these actors. Hiärne advocated the disassociation of the mining business from the influence of keeper entities and trolls, even though he tacitly admitted that they existed. Belief in the possibility of telepathic communication at a distance, and certain forms of magical practices, continued to be widespread even among professed mechanical philosophers and was rarely rejected offhand up until the beginning of the 1760s. They were, though, subject to an increasing degree of skepticism, ridicule, and hostility. Transmutative chymistry, magical practices, and keeper entities were gradually removed from the productive context of mining and smelting. Simultaneously, belief in premonition, telepathy, and subtle, spiritual, or semi-spiritual entities continued in a more private context. Few questioned their ontological status: they continued to form a class of objects and phenomena assumed to exist in nature. But they were nevertheless gradually removed from discourse about nature, which was limited to issues of how nature could be exploited in useful production.

The fate of alchemy differed somewhat from that of the other fields discussed above. Alchemy as a pursuit separate from chemistry was an immediate concern to no more than a few. Therefore it could probably have been discussed, as long as discussions did not venture too far into theology. It was local circumstances, the existence of several powerful alchemists in the higher echelons of the Swedish state, that prevented discussions from reaching the public before the 1760s. In contrast, strange and inexplicable phenomena befell almost everyone, and the issue of whether spirits and angels existed and had agency in this world was a relevant concern for most people. Belief in angels and spirits was of course endorsed by Lutheran orthodoxy, and by the authorities.

This reappraisal of the relationship between religion and natural philosophy among the Swedish eighteenth-century learned elite shows that many in this community embraced what I would call a syncretic worldview, merging folk beliefs with late Cartesianism and Lutheranism.[81] This worldview allowed for the natural occurrence of a number of spiritual phenomena, but also for the gradual fading of belief in phenomena and entities that were strongly associated with material reality. As a result of these circumstances, the early phase of Swedish

Enlightenment debates were polite and low key. Public debate was, of course, also stifled by fear of censorship from religious authorities. It took the "outing" of Swedenborg to make natural philosophers take a more confronting stance. Uninformed as they were of the theological intricacies of his teaching, they instead poised their attacks in a much wider and indiscriminate way. Rather than engaging with Swedenborg's theology, his critics saw his revelations as superstition and used them as ammunition when they joined their voices to the public, enlightened debates already raging elsewhere in Europe.

7

Conclusion: Material Reality and the Enlightenment

In 1808, looking back at the previous decades of revolution and war, Johann Wolfgang von Goethe's friend, the aging Pietist Johann Heinrich Jung-Stilling, passed judgment on the era. Mechanical philosophy was the evil at the root of it all. It had made men believe that everything could be explained through mechanical causes. It had estranged them from a proper belief in the spirit world and had distanced God from his Creation, if only in the minds of men. It had created atheists in great numbers and had made a fashion of skepticism and the ridicule of faith.[1]

It is easy to think that Jung-Stilling was on to something. This book has discussed witches, trolls, angels, premonitions, transmutative chymistry, mechanical philosophy, and utilitarian, patriotic science. We have seen how an environment for transmutative chymistry was established at the Bureau of Mines from the 1680s through the 1720s, and how the Bureau was transformed into a stronghold of mechanical chemistry and mineralogical systematics during the period between 1730 and 1760. It does not overstate the case to say that the knowledge domain of chymistry was transformed at the hands of the Bureau's officials, with far-reaching technical, scientific, and cultural consequences. At the center of the story is the encroachment of mechanical philosophy on the knowledge areas of chymistry and chymical matter theory, as well as chymistry's gradual transformation into mechanical and then mineralogical chemistry: a cameral science at the service of the state.

But is the mechanization of chymistry really at the root of this story? Unlike Jung-Stilling, I do not believe that a complex process of cultural and intellectual change can be reduced to a causal chain showing philosophy's influence on science and culture. The creation of mechanical and mineralogical chemistry cannot be understood without the complementary story of chemists' boundary work, their enclosure of permissible discourse about matter. That story is one of marginalization, as well as of innovative reformulation and recategorization. From our story, we can tease out a number of specific conclusions concerning chemistry's changing roles in relation to mechanical philosophy, to the Enlightenment, to the creation of the category of superstition, to technical and economic reform, and to early modern circulation of knowledge.

Chemistry and Enlightenment

A central theme of this book has been the rejection of a certain range of objects of knowledge, including magic and alchemy, as superstition. The book has outlined how a group of natural philosophers and chymists moved from the open-ended study of curious phenomena to that of ingenious inventions, and from there to a highly utilitarian chemistry. Simultaneously, the new chemists became increasingly skeptical of a range of phenomena, for example, witchcraft, magic, trolls, keeper entities, ghosts, prognostication, telepathy, subtle matters, and claims to direct communication with the spirit world and angels. This change has been described as a slowly unfolding recategorization of entities and phenomena precipitated by changing priorities in knowledge production. But was there really a waning of interest in these phenomena in the eighteenth century? I think not. Rather I would talk about a compartmentalization and, to some extent, cultural displacement. As some phenomena and practices were displaced from the cultural contexts of mining science and productive work, they continued to thrive in others. Distancing was forceful in areas tightly connected to industry and manufacture, but weak in areas associated with the domestic sphere and religion.

Furthermore, the process through which beliefs and truth claims were changed was partial and piecemeal. It was mostly limited to a small group of well-educated men belonging to the middle and upper classes, although not dissimilar processes presumably took place in other social spheres and among other orders of society. It is clear, however, that not even among established, well-educated upper- and middle-class males was there a closure of discussions. And if one ventures outside of those established circles, there is slight trace of a general enlightened intellectual revolution in the eighteenth century.[2]

The separation of alchemy from chemistry is a case in point. It was effected partly through the successful projection of mechanical philosophy onto the world of matter, and the creation of a mechanical and utilitarian chemical discourse about mining and smelting. Mechanical philosophy created bounds around the new knowledge area of chemistry. It set in place rules about proper discourse and actions that were to be followed by actors who wished to be taken seriously.

Of similar importance was the chemists' redefinition of chemical reality as composed of discreet objects, a redefinition that transformed nature into a storeroom for commodities. A major conclusion of this study concerns the social construction of metals as nothing but metallic elements and material commodities. Among historians of science this process has been largely unacknowledged. Even recently, Klein and Spary have described metals as essentially uncontested objects of inquiry in the eighteenth century and have sought to disassociate them from learned discussions about the ultimate structure of matter.[3] In the present study we have seen how the notion of metals as metallic elements followed from the merging of chemistry and natural history with the artisanal practice of assaying. This permitted them to become materials for a chemistry that was concerned almost exclusively with commodification and economic gain.

This new chemistry had little to do with morality, or with an increasingly inaccessible otherworld. Eighteenth-century chemists opted out of metaphysics and were content with describing new substances within the paradigm of chemical composition. We see chemistry emerge as a full-blown cameral science due to these transitions. When the Swedish state turned to chymists, it was as part of an attempt to discipline economic knowledge and to make it reliable, trustworthy, and patriotic. These first attempts to turn chymistry into a cameral science, connected to state bureaucracies, would by and large meet with success in the eighteenth century.

The bureaucratization of chemistry is an important change that highlights not just the intellectual but the social side of chemistry's transformations in the eighteenth century. The demise of seventeenth-century chymistry truly signified the parting of ways of two research communities, their social practices, networks, epistemologies, and truth claims. Transmutative chymistry, or alchemy, was denied a broader institutional base, as well as its scientific status, during the eighteenth century. I would suggest that chymists became chemists when they were embedded in an institutional framework provided by states and academies. As chymists were allocated a space and a budget in eighteenth-century bureaucracies, they changed their priorities and became bureaucrats.

On the other hand, chymists became alchemists when they were preoccupied with knowledge areas—transmutation and chrysopoeia—that could not be fitted into the new institutional frameworks provided for knowledge producers by states and teaching institutions. But there is no doubt that alchemy continued to exist and even thrive throughout the eighteenth century and beyond. It continued to make use of the traditional social settings that it always had used: patronage by royalty and other wealthy notables, and private studies conducted by sufficiently wealthy amateurs. There is no secure indication that the number of alchemical practitioners, or for that matter the cultural power of alchemical symbolism, has been in a slow and steady decline since alchemy parted ways with chemistry. All we can say is that both alchemy as a field of study and the social arenas that it exploited lost in cultural influence and scientific status during the eighteenth century.

The fate of magic is another case that illustrates the problems of describing eighteenth-century intellectual change as a simple linear process. Magic's decline has a measure of linearity to it. As Copenhaver has put it, "the decline of natural magic as a normal and legitimate concern of Western natural philosophy . . . was one of the most important features of the Scientific Revolution."[4] This book bears clear testimony to this gradual decline. Natural magic, however, signifies a specific early modern complex of theories of how the world functions and should be understood. As such it is comparable to, say, Aristotelian matter theory or Paracelsian medicine, which also were in decline during this period. Confusion arises when, as often happens, this or any other early modern theory of magic is conflated with magic in a more late modern sense. In contemporary usage, magic is mostly a catchall category signifying an array of vaguely related practices and beliefs. In the end it is little more than a trash heap for epistemic objects that cannot be located anywhere else. Just like alchemy it has become a derisive Enlightenment term signifying, more or less, not-science neither then nor now. Such use is historically incorrect and belies the fact that intellectual engagement with alchemy's and magic's objects of knowledge has continued to this day.

This study suggests that as an alternative to narratives of decline and general cultural upheaval, it is fruitful to deploy an institutional perspective. Academies, universities, and state administrations grew in size and status during the eighteenth century and played a significant role in the Nordic and German Enlightenment. Simultaneously, new categories and systems of knowledge were developed that were well suited to the forms of life and activities that flourished in these expansive environments. Skepticism, rationalism, utilitarianism, and materialism all had, and still have, their place there. It remains to be proven that Enlightenment sentiments such as philosophical skepticism and rationalism ever had a deep impact on the thinking and habits of the parts of the population that had little connection to these new environments, and the habits of thought and forms of life that came with them.

I would turn the table on all arguments for the decline of this or that phenomena and say that historians of Enlightenment should make a closer study not of decline but of growth. Study instead the emergence and growing cultural influence of institutional structures and social environments such as that of the Bureau of Mines—regulated, bureaucratic places that ever since the eighteenth century have provided work for natural philosophers as long as they have not utilized untamed, suspicious, wild, or deviant theories and knowledge-organizing schemes such as natural magic and chymical gold-making. Such a shift of perspective would bring fresh ideas to the study of the Enlightenment, as well as to the study of changes in eighteenth- and nineteenth-century culture at large.

I would not, however, advocate too much emphasis on institutional structures, paired with a disregard of individual efforts. It was hard and long work, performed both in various public social arenas as well as at mining sites and in laboratories, that created support for the notion that chemical reality was comprehensible in mechanical terms. This work resulted in a partial recirculation of a number of objects of knowledge from nature to imagination. The eighteenth-century explosion of imaginary fiction is well known. An outpouring of novels and romances, aimed at both an upper- and lower-class readership, indicates that many mused about and daydreamed of historical and imaginary realms.[5] Indeed, eighteenth-century imagination expanded with a speed that was equal to the enclosure of socially and scientifically proper knowledge about nature. Can we perceive this as a circulation of knowledge from the realm of nature and external reality to that of imagination and fantasy? Yes, if only in the sense that the Enlightenment's categorization of trolls, witches, ghosts, angels, transmutations, omens, premonitions, and telepathic communication as imaginary has set the stage for how we discuss these phenomena today. These objects of knowledge can be assigned to a common ontological category only when transferred from nature to the realm of imagination and fiction. Their single common denominator is that they are defined as nonentities in nature. Enlightenment categorization casts a long shadow. Even today it is difficult to discuss these phenomena without recourse to categories such as supernatural, parapsychology, or extraordinary phenomena. Somehow late modern minds find it difficult to relate to the knowledge and experiences that lie beneath these categories, without describing them as outside of nature and everyday experience. This is a legacy of Enlightenment redefinitions of nature, as described in this book.

Is this book, then, a chapter of that linear process of change, which Max Weber famously called the disenchantment of nature?[6] Yes and no. Mechanical chemistry's definition of matter as dead and inanimate would achieve a wide cultural influence. Therefore one must grant Weber the concession that this is a story of the disenchantment of matter. It is not, however, a story about the disenchantment of nature—or of the world. Rather, a conclusion of this study is that what gets suppressed in one area seems to pop up in another. The process of historical change only seems to be linear when it is seen from the point of view of enlightened reformers—or their professed opponents. When one adopts other perspectives, there is no discernable waning or fading of interest in, in Weber's sense, enchanted phenomena. Rather we should talk about a constant recompartmentalization and redefinition and, to some extent, cultural displacement, along the lines of Gieryn's notion of boundary work.

There is a more tricky Weberian question, however: the issue of whether the historical processes discussed in this book are best described in Weberian terms as an instance of how the disenchantment of matter interplayed with the emergence of capitalist rationality through the rise of cameralist utilitarianism.[7] I would answer that the story lets itself be described in Weberian terms because Weber was concerned with similar questions, and studied similar groups of actors. But to answer that question in full, it is necessary to discuss the multipolarity of Enlightenment and of eighteenth-century knowledge production.

Circulation of Knowledge in the Eighteenth Century

For the greater part of the eighteenth century the Bureau was widely admired as the cutting edge in cameralist administration, engineering, chemistry, and mineralogy among Europeans, and Britons, in the know. The extent of its influence, however, is largely unknown today, in part due to inherited disciplinary and national division lines. Discipline-oriented historians of eighteenth-century mining, mineralogy, chemistry, and to some extent mechanics all acknowledge an amount of what has been called Swedish influence in their particular fields of study. But prior to this study, it seems that no one really has bothered to read literature across disciplinary and national boundary lines to ascertain its extent. Nationalistic biases in older scholarship may play a part, but they are not a deciding factor.

Rather, I would say that older scholarship has proceeded from an unspoken, perhaps even unconscious, centrist historiographical assumption: that eighteenth-century innovation proceeded from England. Simply put, the Bureau's influence has remained understudied because historians who have come across it have wanted to connect technical and industrial developments on the European continent to the English industrial revolution and have tended to neglect other exchanges. This study balances and gives perspective to such unspoken assumptions, and to the substantial literature on the mechanical and technical aspects of early modern mining.

Although the book presents the information flows of European eighteenth-century chemistry and mining in a way that highlights the importance of the Bureau of Mines, it is pointless to ask to what extent there was a thoroughgoing Swedish influence on German and European developments. This book is not about diffusion but about the interactions among multiple sites of knowledge production. The administrative entity that was the Bureau of Mines was created by immigrants and travelers. Its prominence was in part a direct consequence of its methodical and systematic gathering of information about both domestic and foreign mining regions. To the extent that the Bureau of Mines became a center for knowledge production, it was because it was engaged in an extensive network of information exchange. Its stipendiaries and officials en route between Dresden, Freiberg, Leiden, London, Stockholm, and Uppsala as well as numerous rural Swedish locations were the true centers of the Bureau of Mines knowledge.

Disregarding the notion of centers and peripheries, we can look at the role of chymistry and chemistry in late seventeenth- and eighteenth-century mining with fresh eyes. All over Europe powerful state actors, driven by economic and political concerns, instigated changes and reforms in mining, smelting, industry, and manufacture. As we have seen, chemistry played an important role for changes in work procedures, technological change in especially smelting, and for the eighteenth century's reinterpretation of mining as a scientific and rational enterprise. Older scholarship has tended to emphasize the importance of Freiberg in the emergence of scientific mining, as evident in a new combination of theory and practice in chemical mineralogy, natural history, physics, subterranean geography, mine surveying, mechanics, et cetera at the Freiberg mining school from 1765.[8]

This study establishes that this assemblage of knowledges and skills was not created in Freiberg at that time. Not only did these these knowledges and skills have a long and successful pre-eighteenth-century history. Arguably, it was the officials of the Bureau of Mines—in continuous interaction with various European counterparts—who rebranded many of these pursuits as cameral mining sciences during the years 1730–60.

Let us return now, for the last time, to the issue of matter as the building blocks of material reality. This study locates a radical break in chemical theory and practice to the first half of the eighteenth century, and to natural philosophers active at the Bureau. The Bureau's officials were important for the creation of the modern notion of matter as solid and composed out of elements: that is, well-defined building blocks accessible through chemical analysis, and conceived of as the constituent parts of the objects of the natural world. As this notion gained a wider hold throughout Europe, it contributed to the displacement of notions of matter as permeable, unstable, and subject to spiritual influences and impressions. When matter was redefined as solid and as made from useful components, the influence of less material forces and entities, such as spirits and transmutations, was increasingly defined as useless speculation and hardly worth considering for serious chemical practitioners. In this way, the preoccupation of the Bureau's chemists with composition fed into and supported Enlightenment criticism of a range of nonchemical phenomena, which increasingly came to be classed as outside of nature and nonexistent. From this point of view, what we have studied here is indeed an episode in the mechanization of the Western worldview.

Acknowledgments

During most of the time spent writing this book, I have
worked as an assistant professor at the Division for History of Science and Technology at the Royal Institute of
Technology (KTH) Stockholm. I extend a heartfelt thank
you to all those who were connected to this environment
during my time there. There are many who have made a
special effort on my behalf by reading the manuscript and
giving me valuable advice on specific points. I particularly
thank Bjørn Ivar Berg, Maria Berggren, David Dunér,
Matthew Eddy, Eva Fee (thanks, mom!), Anders Houltz,
Anders Lundgren, the late Marie Nisser, Linda Oja, Jacob
Orrje, Mathias Persson, Theodore M. Porter, Lawrence
Principe, Jenny Rampling, Göran Rydén, Kristiina Savin,
Karin Sennefelt, and Otto Sibum. I also thank those who
gave me valuable input at conferences and seminars in Los
Angeles and Philadelphia (United States), Estoril (Portu-
gal), Steinhaus/Cadipietra (Italy), Louvain/Leuven (Bel-
gium), Freiberg (Germany), Oxford (England), Turku/
Åbo (Finland), and Stockholm and Uppsala (Sweden). I
extend special thanks to Svante Lindqvist for the gift of
his collection of literature on early modern mining. Another special thanks is due the director Karl Grandin and
the archivists Maria Asp and Anne Miche de Malleray at
the Center for History of Science at the Royal Swedish
Academy of Sciences, where I spent two months as a re-

searcher in 2010. My work would have been impossible without the aid of archivists and librarians at the National Library of Sweden, the National Archives of Sweden, and the university libraries of Uppsala and the Royal Institute of Technology, Stockholm (and in particular Tommy Westergren, at the latter). Thanks also to the Clark Library and the Huntington Library in Los Angeles and the Chemical Heritage Foundation, Philadelphia, as well as the University Library of Rostock. Finally thank you to the late Gary Gygax, whom I never had the pleasure to meet or talk with. As a fifteen-year-old I would hardly have imagined myself still considering the distinctions between trolls and kobolder at age forty-two, although I would have been pleased had I known.

The research for this book was funded by Vetenskapsrådet (the Swedish science council) as a three-year independent researcher grant. I was also very much helped to finish it by a short-term research position granted me by my present employer, the Department for History of Science and Ideas, and the Office for the History of Science at Uppsala. Additional grants have been generously provided by the Swedish Royal Academy of Sciences (from Stiftelsen Nordenskjöldska Swedenborgsfonden), the Magn. Bergwalls Foundation, the Helge Ax:son Johnson Foundation, Jernkontoret (the Swedish Steel Producers' Association), and the Swedish Chemical Society. I have also benefited from the hospitality of the Center for Seventeenth and Eighteenth Century Studies at UCLA, which granted me the status of visiting scholar during my stay in Los Angeles during the academic year 2007–8.

Notes

CHAPTER 1

1. Historians have made heterogenous use of the term *the Enlightenment*. Some see it as a pan-European or global phenomena. Others have discussed it as a distinctly French intellectual movement, connected to the circles around Voltaire and *L'encyclopédie*. Here Enlightenment is discussed in relation to knowledge production in natural philosophy from about 1680 to 1760. It is mainly used as an actor's category. Hence it does not denote a movement or a historical epoch but an ideal and an object of identification. Many in the eighteenth century used the word to signal their general support of intellectual clarity, of scientific, economic, and cultural progress, as well as their intent to disassociate themselves from what they perceived as the irrationalities of the recent past. An analogy is found in the way the term *modernity* was used in the twentieth century. For some representative positions in the debate on Enlightenment, see Roy Porter and Mikuláš Teich, *The Enlightenment in national context* (Cambridge: Cambridge University Press, 1981); Roy Porter, *The Enlightenment* (New York: Macmillan, 1990); and Dorinda Outram, *The Enlightenment* (Cambridge: Cambridge University Press, 1995). Alan Charles Kors, "Preface," in A. C. Kors, ed. in chief, *Encyclopedia of the Enlightenment* (Oxford: Oxford University Press, 2003), 1:xvii–xxii; Lynn Hunt and Margaret Jacob, "Enlightenment studies," ibid., 1:418–30. Jonathan Israel, *Radical enlightenment: Philosophy and the making of modernity, 1650–1750* (Oxford: Oxford University Press, 2001). J. Israel, *Enlighten-*

ment contested: Philosophy, modernity and the emancipation of man, 1620–1752 (Oxford: Oxford University Press, 2006).

2. Body parts from many of these beings had been prominently displayed in the curiosity cabinets of the sixteenth and seventeenth centuries but were discredited and no longer considered real naturalia during the eighteenth. Paula Findlen, "Inventing nature: Commerce, art and science in the early modern cabinet of curiosities," in P. H. Smith and P. Findlen, *Merchants and marvels: Commerce, science, and art in early modern Europe*, 297–323 (New York: Routledge, 2002), 307–20.

3. On chemistry's position as the premier science of materials, Ursula Klein and Wolfgang Lefèvre, *Materials in eighteenth-century science: A historical ontology* (Cambridge: MIT Press, 2007), 1–2, 8–9.

4. Cyril Stanley Smith, "The texture of matter as viewed by artisan, philosopher, and scientist in the seventeenth and eighteenth centuries," in Cyril Stanley Smith and John G. Burke, eds., *Atoms, blacksmiths, and crystals: Practical and theoretical views of the structure of matter in the seventeenth and eighteenth centuries* (Los Angeles: William Andrews Clark Memorial Library, 1967), 3–34; quotation on 4. More recently economic historian Göran Rydén remarked that "the relationship between eighteenth-century thought and the material world of making and selling commodities is not a conspicuous feature of recent Enlightenment scholarship." Göran Rydén, "The Enlightenment in practice: Swedish travellers and knowledge about the metal trades," *Sjuttonhundratal* (2013): 1. Hence, there is a need of analyses that reconcile narratives of the Enlightenment with those of the Industrial Revolution. Ibid., 8.

5. Ian Bostridge, *Witchcraft and its transformations c. 1650–c. 1750* (Oxford: Clarendon Press, 1997), 2–3, 234, 240–43. Robert M. Burns, *The great debate on miracles: From Joseph Glanvill to David Hume* (Lewisburg, PA: Bucknell University Press, 1981), 9–12. Stuart Clark, *Thinking with demons: The idea of witchcraft in early modern Europe* (Oxford: Clarendon Press, 1997), 150–215; Lorraine Daston and Katharine Park, *Wonders and the order of nature: 1150–1750* (New York: Zone Books, 1998), 10–14, 126–27, 250–51, 331–39.

6. Lorraine Daston, "Preternatural philosophy," in Lorraine Daston, ed., *Biographies of scientific objects*, 15–41 (Chicago: University of Chicago Press, 2000), 37. Daston and Park, *Wonders and the order of nature*, 126–27.

7. Daston, "Preternatural philosophy," 37.

8. This was of course a complex process, also involving commercial aspects not discussed here. See further, Benjamin Schmidt, "Inventing exoticism: The project of Dutch geography and the marketing of the world, circa 1700," in Smith and Findlen, *Merchants and marvels*, 347–69. Liliane Pérez, "Technology, curiosity and utility in France and England in the eighteenth century," in B. Bensaude-Vincent and C. Blondel, eds., *Science and spectacle in the European Enlightenment* (Aldershot: Ashgate, 2008), 25–42.

9. Due to these considerations I would reject Daston's use of the term preternatural. The preternatural, argues Daston, should be regarded as a specific class of phenomena, that today are generally regarded as supernatural but that were seen as natural by most early modern Europeans. However, to introduce

such a term to signify a wide array of objects and practices that eventually came to be discarded with the formation of modern natural knowledge, is to tacitly lend one's support to a classic enlightenment narrative of progress. Daston, "Preternatural philosophy," 15–41, esp. 37. Daston and Park, *Wonders and the order of nature*, 126–27.

10. This ethos or strain of philosophy is sometimes called utilism. It should not be confused with the twentieth-century utilitarian school of philosophy. It is useful to consider it as the common ground that united what Jonathan Israel has called the philosophies of the radical Enlightenment and the moderate mainstream Enlightenment. Israel, *Enlightenment contested*, 10–11. See also Karin Johannisson, "Naturvetenskap på reträtt: En diskussion om naturvetenskapens status under svenskt 1700-tal," *Lychnos* (1979–80); Lisbet Koerner, "*Daedalus Hyperboreus*: Baltic natural history and mineralogy in the Enlightenment," in W. Clark, J. Golinski, and S. Schaffer, eds., *The sciences in Enlightened Europe* (Chicago: University of Chicago Press, 1999), 389.

11. William Newman and Lawrence Principe, "Alchemy vs. chemistry: The etymological origins of a historiographic mistake," *Early Science and Medicine* no. 3, 1(1998): 38–41; Principe, *The aspiring adept: Robert Boyle and his alchemical quest* (Princeton: Princeton University Press, 1998), 8–10. For recent positions of these authors, William Newman, "What have we learned from the recent historiography of alchemy?" *Isis* 102 (June 2011): 313–21. Lawrence Principe, "Alchemy restored," *Isis* 102 (June 2011): 305–12.

12. Focus of discussion in this book is on chymistry/chemistry/alchemy, keeper entities/trolls, extrasensory perception, spirits, angels, devils, and matter theory. I make only cursory incursions into the adjacent fields of land and mine surveying, treasure hunting, and dowsing. For a discussion of these areas, see Warren Dym, *Divining science: Treasure hunting and earth science in early modern Germany* (Leiden: Brill, 2011).

13. D 6:3 Bergskollegiums arkiv, huvudarkivet. Axel Fredrik Cronstedt, *Bref til Hr . . . om den mystiska naturkunnogheten Stockholm den 20 Maj 1758*, 91. RA.

14. Thomas F. Gieryn, *Cultural boundaries of science: Credibility on the line* (Chicago: University of Chicago Press, 1999), 5.

15. Ibid., xi, 1–6; quotation on xi.

16. For this approach, see Betty Jo Teeter Dobbs, *The foundations of Newton's alchemy, or, "The hunting of the greene lyon"* (Cambridge: Cambridge University Press, 1975); Steven Shapin and Simon Schaffer, *Leviathan and the airpump: Hobbes, Boyle, and the experimental life* (Princeton: Princeton University Press, 1985); Principe, *Aspiring adept*; Martin Rudwick, *The great Devonian controversy: The shaping of scientific knowledge among gentlemanly specialists* (Chicago: University of Chicago Press, 1985).

17. For a case in point, Martha Chaiklin, "Simian amphibians: The mermaid trade in early modern Japan," in Yoko Nagazumi, ed., *Large and broad: The Dutch impact on early modern Asia: Essays in honor of Leonard Blussé* (Tokyo: Toyo Bunko, 2010), 241–73.

18. Ursula Klein and E. C. Spary, "Introduction: Why materials?" in U. Klein and E. C. Spary, *Materials and expertise in early modern Europe: Between*

market and laboratory (Chicago: University of Chicago Press, 2010), 1, 4–5, 8–10; quotations on 1, 4–5.

19. Ibid., 4.

20. On their definition of hybrid expert, see ibid., 6. The problems with this approach are evident from the hopefully unintended whiggish anachronism of passages such as "[a] sixteenth-century alchemist ordered commodities such as metals into a web of signs." Here the historical actor's interpretation of metals through theories of correspondence is clearly not assigned the same status of truth as the property of commodity assigned to metals by Klein and Spary. Ibid., 11.

21. Kapil Raj, *Relocating modern science: Circulation and the construction of knowledge in South Asia and Europe, 1650–1900* (Houndmills, Basingstoke: Palgrave Macmillan, 2007), esp. 13–14, 18–21. The term *circulation* is preferable to the term *transfer* as the latter contributes to the construction of an implicit knowledge hierarchy between the place of origin of the transfer and the place of reception. Hence, although the concept of transfer by no means gives an incorrect description of certain events, the preponderance of case studies of transfers draws undue attention to unequal distributions of knowledge by the force of the examples they provide. Cf. Svante Lindqvist, *Technology on trial: The introduction of steam power technology into Sweden, 1715–1736* (Uppsala: Almqvist and Wiksell, 1984), 11–12.

22. The pursuit of science or philosophy was never a primary concern of the Bureau as a whole, however. Its primary tasks were to propose legislation for the mining industry and to serve as an appeals court, grant privileges for new mining enterprises, control the quality of mining produce, and improve the business through technological innovations and other means.

23. The beginning of the incursion of scientific ideas and modern technologies into the mining business has often been located in the second half of the eighteenth century. Particular emphasis has been laid on the founding of mining academies and the teaching of mining sciences at universities. The Bureau of Mines, although a teaching institution, and its great influence during the first half of the eighteenth century tend to be neglected in such narratives. This is true of both Swedish and international historiography. See Björn Ivar Berg, "Das Bergseminar in Kongsberg in Norwegen (1757–1814)," *Der Anschnitt: Zeitschrift für Kunst und Kultur im Bergbau* 3–4 (2008): 154–55. Michael Fessner, "Die Berliner Bergakademie von ihrer Gründung bis zur Eingliederung in die Technische Hochschule Berlin-Charlottenburg (1770–1916), in A. Westermann and E. Westermann, eds., *Wirtschaftslenkende Montanverwaltung- Fürstlicher Unternehmer—Merkantilismus: Zusammenhänge zwischen der Ausbildung einer fachkompetenten Beamtenschaft in der staatlichen Geld- und Wirtschaftspolitik in der Frühen Neuzeit* (Husum: Matthiesen Verlag, 2009), 439–43. Ursula Klein, "The Prussian mining official Alexander von Humboldt," *Annals of Science* 69, no. 1: 29–31, 33–34, 60–62. Peter Konečný, "The hybrid expert in the 'Bergstaat': Anton von Ruprecht as a professor of chemistry and mining and as a mining official, 1779–1814," *Annals of Science* 69, no. 3 (July 2012): 335, 346. On the lack of research on the Bureau of Mines, see Lindqvist, *Technology on trial*, 95–96. Colin A. Russell, "Science on the fringe of Europe: Eighteenth-

century Sweden," in D. Goodman and C. Russell, eds., *The rise of scientific Europe, 1500–1800* (London: Hodder and Stoughton, 1991), 305–32. Maths Isacson, "Bergskollegium och den tidigindustriella järnhanteringen," *Daedalus: Tekniska museets årsbok* årg. 66 (1998): 43–58. On Swedish iron production, Karl-Gustaf Hildebrand, *Swedish iron in the seventeenth and eighteenth centuries: Export industry before the industrialization* (Stockholm, 1992); K.-G. Hildebrand, "Gammalt och nytt i det svenska järnets historia: En översikt över fem årtionden," *Daedalus: Tekniska museets årsbok* 65 (1997): 1–30; Chris Evans and Göran Rydén, *Baltic iron in the Atlantic world in the eighteenth century* (Boston: Brill, 2007), 31–37.

24. Simon Werrett, "An odd sort of exhibition: The St. Petersburg Academy of Sciences in enlightened Russia" (diss., University of Cambridge, March 2000), 9. Abbri has put forward a similar argument addressing the historiography of the so-called Chemical Revolution. Fernando Abbri, "Some ingenious systems: Lavoisier and the northern chemistry," in M. Beretta, ed., *Lavoisier in perspective*, 95–108 (Munich: Deutsches Museum, 2005), 96–97.

25. Werrett, "Odd sort of exhibition," 9, 12; quotation on 12.

26. Sverker Sörlin, *De lärdas republik: Om vetenskapens internationella tendenser* (Malmö: Liber-Hermod, 1994), 42.

27. For some recent works that seek to amend this old fallacy, Jonathan Simon, *Chemistry, pharmacy and revolution in France, 1777–1809* (Aldershot, UK: Ashgate, 2005); Tara Nummedal, *Alchemy and authority in the Holy Roman Empire* (Chicago: University of Chicago Press, 2007). Matthew D. Eddy, *The language of mineralogy: John Walker, chemistry and the Edinburgh Medical School, 1750–1800* (Farnham, UK: Ashgate, 2008).

28. Two texts stand out among discussions of the international impact of Bureau of Mines chemistry: Theodore M. Porter, "The promotion of mining and the advancement of science: The chemical revolution in mineralogy," *Annals of Science* 38 (1981): 549–51, 558–60. Russell, "Science on the fringe of Europe," esp. 323. In contrast, the Bureau of Mines chemists (with the exception of Sven Rinman) were something of a blind spot to Cyril Stanley Smith. Cyril Stanley Smith, "The interaction of science and practice in the history of metallurgy," *Technology and Culture* 2, no. 4 (1961): 359–61; and C. Smith, "The discovery of carbon in steel," *Technology and Culture* 5, no. 2 (1964): 149–75.

29. Indeed, although Sweden has been assigned an important place in the overall history of eighteenth-century chemistry, focus tends to be put on the period after 1750, and the work of Johan Gottschalk Wallerius, Torbern Bergman, and Carl Wilhelm Scheele. Henry Guerlac, "Some French antecedents of the chemical revolution," *Chymia: Annual Studies in the History of Chemistry* 5 (1959); Heinz Cassebaum and George B. Kauffman, "The analytical concept of a chemical element in the work of Bergman and Scheele," *Annals of Science* 33 (1976); Ursula Klein, "The chemical workshop tradition and the experimental practice: Discontinuities within continuities," *Science in Context* 9 (1996); Christoph Meinel, "Theory or practice? The eighteenth-century debate on the scientific status of chemistry," *Ambix* (1983); Evan Melhado, "Mineralogy and the autonomy of chemistry around 1800," *Lychnos* 1990; Lissa Rob-

erts, "Filling the space of possibilities: Eighteenth-century chemistry's transition from art to science," *Science in Context* 6 (1993); Mi Gyung Kim, "Lavoisier, the father of modern chemistry?" in M. Beretta, ed., *Lavoisier in perspective*, 167–91 (Munich: Deutsches Museum, 2005). For more general overviews, J. Golinski, "Chemistry," in R. Porter, ed., *The Cambridge history of science*, vol. 4: *Eighteenth-century science* (Cambridge: Cambridge University Press, 2003), 391; Frederic L. Holmes, "Chemistry," in Kors, *Encyclopedia of the Enlightenment*, 1:228.

30. The period ca. 1680–1760 also covers the heyday of the Bureau as an environment for production of knowledge. From the 1680s the Bureau began to employ *curiosi* and started to develop an internal system of education. The far end of the study is placed at 1760. At this time there had emerged a number of environments that would eventually replace the Bureau. These were the Swedish Royal Academy of Sciences (1739), the Swedish Ironmasters' Association (Jernkontoret, 1747), and finally the Uppsala chair of chemistry (1750). In particular Sweden's first professor of chemistry, Johan Gottschalk Wallerius, soon managed to turn Uppsala (rather than the Bureau of Mines) into the main venue for chemistry in Sweden. This process of relocation would be continued by Wallerius's successor Torbern Bergman. Hjalmar Fors, *Mutual favours: The social and scientific practice of eighteenth-century Swedish chemistry* (Uppsala: Uppsala Universitet, 2003); H. Fors, "J. G. Wallerius and the laboratory of Enlightenment," in E. Baraldi, H. Fors, and A. Houltz, eds., *Taking place: The spatial contexts of science, technology and business*, 3–33 (Sagamore Beach: Science History Publications, 2006). See also Sven Widmalm, "Instituting science in Sweden," in R. Porter and M. Teich, eds., *The scientific revolution in national context*, 240–62 (Cambridge: Cambridge University Press, 1992).

31. The passage above is no more than an interpretation of the so-called *strong program of the sociology of knowledge* as formulated by David Bloor, *Knowledge and social imagery*, 2nd ed. (1976; Chicago: University of Chicago Press, 1991), 7. For a discussion of early modern mining practices with many parallels to the present study, see Dym, *Divining science*, 3, 21.

32. Steven Shapin, *A social history of truth: Civility and science in seventeenth-century England* (Chicago: University of Chicago Press, 1994), 20–22, 35–38; quotation on 36.

33. The term *troll* sometimes was used by actors as a generic term that included not only keeper entities but also devils, giants, fairies, etc. Mikael Häll, *Skogsrået, näcken och djävulen: Erotiska naturväsen och demonisk sexualitet i 1600-och 1700-talets Sverige* (Stockholm: Malörts förlag, 2013), 78, 86.

34. For Germany, see Christoph Bartels, *Vom früneuzeitliche Montangewerbe zur Bergbauindustrie: Erzgebau im Oberharz 1635–1866* (Bochum: Deutsches Bergbau-Museum, 1992), 85–86.

CHAPTER 2

1. "Kongl. Commissorial Rättens Ransakning om Trollväsendet i Stockholm åren 1676. 1677.," in [C. Adlersparre], ed., *Historiska samlingar*, 5:214–

383 (Stockholm: F. B. Nestius, 1822), 216–17. An authoritative overview of the Swedish witchcraft trials can be found in Bengt Ankarloo, *Trolldoms-processerna i Sverige*, 2nd ed., Rättshistoriskt bibliotek 17 (Stockholm: Nordiska Bokhandeln, 1984). On the Stockholm trials, ibid., 194–214. For a complementary (and more recent) perspective, Per Sörlin, *"Wicked arts": Witchcraft and magic trials in southern Sweden, 1635–1754* (Leiden: Brill, 1999).

2. On Hiärne and the Royal Society, Sven Rydberg, *Svenska studieresor till England under frihetstiden* (Uppsala: Almqvist och Wiksell, 1951), 66–72.

3. "Kongl. Commissorial Rättens Ransakning," 233, 237, 245–46, 254–55, 266. "Betänkande af Kyrkoherden i St. Jacobs församling Magnus Pontinus om Trolldoms väsendet," in [Adlersparre], *Historiska samlingar*, 5:387–96; dated 6 May 1676, 389–90.

4. "Kongl. Commissorial Rättens Ransakning," 258–59; see also 297–98.

5. Noraeus kept to this vow throughout the trial. As the commission's attitude became markedly more critical, it was not long until most of the witnesses took back their stories. "Kongl. Commissorial Rättens Ransakning," 260–61, 267–84. See also the published testimony of the thirteen-year-old Johan Johansson Grijs, "Poiken ifrån Giefle Johan Johansson Grijs, på sin 13 åhrs ålder, bekiende d. 23. Augusti A:o 1676 i fängelset således.," in [Adlersparre], *Historiska samlingar*, 5:402–7.

6. Urban Hiärne, "Kort betenckiande öffwer de anfechtade barnens i Stockholm förrige klagans återkallelse och den nya bekennelsen inför den Kongl. Commissorial Rätten i Stockholms Stadzhuus som begyntes den 11 Sept:s Anno 1676, och intill dato continuerat" (formal report to Kongl. Commissorial Rätten, of 3 Oct. 1676), in [Adlersparre], *Historiska samlingar*, 5:409–32. See also Hiärne's autobiography, Urban Hiärne, "Salig Landzhöfding Urban Hiernes Lefwernesbeskrifning af honom sielf i pennan fattad," in *Äldre svenska biografier*, ed. H. Schück, 5:139–79, Uppsala Universitets Årsskrift 1916 program 4:2 (Uppsala: Akademiska Bokhandeln, 1916).

7. "Kongl. Commissorial Rättens Ransakning," 316–17; quotations on 281, 293.

8. Ibid., 350, 356–62, 366; quotation on 350.

9. Ibid., 376–77; quotation on 376–77.

10. Sten Lindroth, "Hiärne, Urban," in *Svenskt biografiskt lexikon*, 19:141–50.

11. Odhelius to Urban Hiärne, Brussels, 27 Dec. 1686, in Carl Christoffer Gjörwell, *Det svenska biblioteket* (Stockholm: Wildiska tryckeriet, 1758), 2:303.

12. On the early modern view of the Devil as master deceiver, limited by the laws of natural causation, see Clark, *Thinking with demons*, 166–67.

13. Hiärne, "Kort betenckiande öffwer de anfechtade barnens i Stockholm," in [Adlersparre], *Historiska samlingar*, 5:422, 430, 432.

14. The following reflects the views of magic of influential European theorists. For a sample of Swedish academic discourse on magic and witchcraft, see Ankarloo, *Trolldomsprocesserna i Sverige*, 100–112. The term *magic* was rarely used in early modern Sweden. For general definitions of acts of magic, see Linda Oja, *Varken Gud eller natur: Synen på magi i 1600- och 1700-talets Sverige*

(Stockholm/Stehag: Symposion, 1999), 22. For distinctions between different types of magic in courtrooms (i.e., at the Göta High Court), see Sörlin, *"Wicked arts,"* 27–37.

15. Lynn Thorndike, *A history of magic and experimental science* (New York: Columbia University Press, 1958), 7:273. Clark, *Thinking with demons*, 226, and many other places.

16. Gabriel Naudaeus, *The history of magick by way of apology, for all the wise men who have unjustly been reputed magicians, from the creation, to the present Age. Written in French, by G. Naudaeus Late Library-Keeper to Cardinal Mazarin*, trans. J. Davies (1625; London: John Streater, 1657), 13–18, 26–45, 188. See also Thorndike, *History of magic and experimental science*, 7:302.

17. Naudaeus, *History of magick*, quotations on 22, 35.

18. Ibid., quotations on 37.

19. Ibid., 165–85.

20. Ibid., 186–87; quotation on 187. In fact, not even Cornelius Agrippa was considered a magician according to Naudé's definition. Ibid., 188–201.

21. Michael Hunter, introduction to *The occult laboratory: Magic, science and second sight in late seventeenth-century Scotland. A new edition of Robert Kirk's The Secret Commonwealth and other texts, with an introductory essay by Michael Hunter*, ed. M. Hunter (Woodbridge: Boydell Press, 2001), 1, 8–9, 29–31.

22. During antiquity, philosophers had put forward two major theories to explain the structure of matter. These were the Aristotelian theory of the four elements (that is, fire, air, water, and earth) and atomism. According to the atomism of the ancients, all matter consisted of particles that interlaced randomly in empty space. During late antiquity and the medieval period, this theory was more or less forgotten in western Europe. It was taken up again in the seventeenth century and reformulated by, among others, René Descartes.

23. On Descartes's time in Stockholm, Susanna Åkerman, *Fenixelden: Drottning Kristina som alkemist* (Möklinta: Gidlunds, 2013), 77–89.

24. Rolf Lindborg, *Descartes i Uppsala: Striderna om 'Nya Filosofien' 1663–1689* (Stockholm: Almqvist och Wiksell, 1965), 115–23. R. Lindborg, "Urban Hiärne såsom studerande för medicinprofessorn Petrus Hoffwenius i Uppsala," in S. Ö. Ohlson and S. Tomingas-Joandi, eds., *Den otidsenlige Urban Hiärne: Föredrag från det internationella Hiärne-symposiet i Saadjärve, 31 augusti–4 september 2005* (Tartu: Trükikoda Greif, 2008), 45–50.

25. Widmalm, "Instituting science in Sweden," 241. Lindborg, "Urban Hiärne såsom studerande," 46–49. For an overview of the battles, Gunnar Eriksson, *Rudbeck 1630–1702: Liv, lärdom, dröm i barockens Sverige* (Stockholm: Atlantis, 2002), 129–57.

26. The dissertation was the third part of a treatise authored by Hoffwenius. The corollary is usually attributed to Hiärne, but it could also have been written by Hoffwenius himself. Lindroth implies that Hiärne did not agree with the content of *Mantissa physica*. This is unlikely. In other contexts Lindroth—a positivist admirer of the Enlightenment—allowed his philosophical preferences to severely distort his reading of Hiärne. See, e.g., Sten Lindroth, "De stora häxprocesserna," in J. Cornell et al., eds., *Den svenska historien 7:*

Karl X Gustav, Karl XI. Krig och reduktion (Stockholm: Bonniers, 1978), 159. Many have criticized Lindroth's interpretation of Hiärne; see Lindborg, "Urban Hiärne såsom studerande," 45–50, and Kristiina Savin, "Gud i Naturens laboratorium: Varsel och järtecken hos Urban Hiärne," 116–17, 128, in the same volume.

27. Hiärne, "Salig Landzhöfding Urban Hiernes Lefwernesbeskrifning," 148–60.

28. Ibid., 157. In Hiärne's published autobiography the Italian's name is given as Staniani. The name Praciani above is given as in X268, UUB. The latter text is, according to Strandberg, a far better copy of Hiärne's original manuscript (now lost) than that used by the editors of the published autobiography. Olof Strandberg, *Urban Hiärnes ungdom och diktning* (Stockholm: Almqvist och Wiksell, 1942), 141.

29. As Magnus von Platen put it, "Hiärne came with the inheritance and became the personal physician of both." Bielke now became Hiärne's main patron. Hiärne also received the patronage of Councilor of the Realm Claes Fleming. There was, however, a downside to having older patrons. In 1684–85, both Bielke and Fleming died, and again Hiärne stood without a powerful personal helper. Magnus von Platen, "Den sörjande klienten," in M. von Platen, ed., *Klient och patron: Befordringsvägar och ståndscirkulation i det gamla Sverige*, 51–63 (Stockholm: Natur och Kultur, 1988),54–55; quotation on 55.

30. Lindroth, "Hiärne, Urban," 142.

31. Lindroth has pointed out that there are numerous passages critical of Cartesianism in Hiärne's manuscripts. Sten Lindroth, "Urban Hiärne och Laboratorium Chymicum," *Lychnos* (1946–47): 96.

32. Urban Hiärne, *Defensionis paracelsicae prodromus: Eller kort föremäle af then uthförligare förswars skrift för den stora philosophus theutonicus Theophrastus Paracelsus, som nyligen medelst en hård beskyllning uthan någon orsak är worden antastad* (Stockholm: Julius G. Matthiae, 1709), 12.

33. Anne Conway, one of the more influential late seventeenth-century Neoplatonists, proposed that the process of creation happened as God diminished his light to make room for his creation. Anne Conway, *The principles of the most ancient and modern philosophy* (London, 1692), 5–7. See also Paracelsus, "Tractatus IV. Von dem underscheit der corporum und spirituum," in *Theophrast von Hohenheim gen. Paracelsus, Sämtliche Werke 1. Abteilung Medizinische naturwissenschaftliche und philosophische Schriften*, ed. Karl Sudhoff (Munich, 1931), pt. 13, 350–51. On Paracelsus's influence in the Protestant world, Thomas Karlsson, *Götisk kabbala och runisk alkemi: Johannes Bureus och den götiska esoterismen* (Stockholm: Stockholms Universitet, 2010), 94–98.

34. Åsa Karlsson, "Nyen," in *Karolinska förbundets årsbok 1999, Att illustrera stormakten: Den svenska Fortifikationens bilder 1654–1719* (Lund: Historiska Media, 2001), 60–61.

35. Strandberg, *Urban Hiärnes ungdom och diktning*, 7–18. Piret Lotman, "Kyrkoherdens son från Nevastranden: Föräldrar och barndomsmiljö," in S. Ö. Ohlson and S. Tomingas-Joandi, eds., *Den otidsenlige Urban Hiärne*, 25–33,

and Katrin Askegren, "Provinsen Ingermanland under Hiärnes tid–en kort historik," 35–43, ibid.

36. Strandberg, *Urban Hiärnes ungdom och diktning*, 16–18. On the father's prophecy, Hiärne, "Salig Landzhöfding Urban Hiernes Lefwernesbeskrifning," 142.

37. Hiärne, "Salig Landzhöfding Urban Hiernes Lefwernesbeskrifning," 140–43. Lotman, "Kyrkoherdens son från Nevastranden29–30.

38. Hiärne, "Salig Landzhöfding Urban Hiernes Lefwernesbeskrifning," 144–46. Hiärne was an artistic talent who would finance his studies in part by painting portraits. He also participated in Uppsala's literary circles as a poet, playwright, and novelist. Lindroth "Hiärne, Urban," 142–44. His studies, however, were made possible through the support of his brother, Thomas Hiärne. Hiärne, "Salig Landzhöfding Urban Hiernes Lefwernesbeskrifning," 146.

39. Hiärne, "Salig Landzhöfding Urban Hiernes Lefwernesbeskrifning," 162–63. See also Tilas, "Anteckningar ur Hiärne Ättens Minne," 439–40.

40. Savin, "Gud i Naturens laboratorium," 117.

41. Ankarloo, *Trolldomsprocesserna i Sverige*, 57–67, 91–92, 94–95. Sörlin, *"Wicked arts,"* 18, 70–76.

42. Ankarloo, *Trolldomsprocesserna i Sverige*, 51. For more cases of learned Swedish magicians, ibid., 51–52.

43. Johan Bure, "Anteckningar af Johannes Thomae Agrivillensis Bureus," *Samlaren: Tidskrift utgiven af Svenska literatursällskapets arbetsutskott* 4 (1883): 12–43, 71–126; quotation on 74–75.

44. For a somewhat different interpretation of this episode, see Karlsson, *Götisk kabbala*, 107–8, 113.

45. Hiärne, *Defensionis paracelsicae prodromus*, 6, 9, 12; quotations on 12.

46. That is not to say that such beliefs were unusual. On dreams, see Kristiina Savin, *Fortunas klädnader: Lycka, olycka och risk i det tidigmoderna Sverige* (Lund: Sekel, 2011), 50, 52, 77.

47. Hiärne, "Anhang Betreffande Spådomar och Underwärck," in *Defensionis paracelsicae prodromus*, 18.

48. Paracelsus, "Tractatus V. Von dem schlaf und wachen der leiber und geister," in *Theophrast von Hohenheim gen. Paracelsus, Sämtliche Werke 1*, pt. 13, 352–58.

49. In the Swedish pious literature of the seventeenth century, trolls and similar beings were held to belong to the demonic spirit-world. David Lindquist, *Studier i den svenska andaktslitteraturen under stormaktstidevarvet: Med särskild hänsyn till bön-, tröste-, och nattvardsböcker* (Uppsala: Almqvist och Wiksell, 1939), 361–72. On angels, Savin, *Fortunas klädnader*, 99–103, 114–26.

50. In a paragraph on defamation in a thirteenth-century Swedish code of laws (*äldre västgötalagen*), it was implied that a witch could transform herself into a troll, while simultaneously retaining her own outward appearance. The affinity of the beings come out clearly in the Swedish language. In Swedish, *troll* as a singular noun is used to designate a group of nonhuman sentient beings. But the term was also used for humans. Hence witches were called troll-women (*trollkonor*). Warlocks and wizards (and today, stage magicians) were called troll-men (*trollkarlar*).

51. Urban Hiärne, "Anhang Betreffande Spådomar och Underwärck," 15–18; quotation on 17–18.

52. Paracelsus, "Liber de nymphis, sylphis, pygmaeis et salamandris et de caeteris spiritibus Theophrasti Hohenheimensis. Prologus," in *Theophrast von Hohenheim gen. Paracelsus, Sämtliche Werke 1*, pt. 14, 149–50.

53. Hiärne, "Anhang Betreffande Spådomar och Underwärck," 18. Savin has emphasized the importance of Hiärne's concept of *syndrome signorum*, or juxtaposition of several signs, according to which a single sign means nothing, whereas several signs together indicate that important events are in the making. Savin, "Gud i Naturens laboratorium," 127.

54. Olaus Magnus, *Historia om de nordiska folken: Deras olika förhållanden och villkor, plägseder, religiösa och vidskepliga bruk, färdigheter och idrotter, samhällsskick och lefnadssätt, krig, byggnader, och redskap, grufvor och bergverk, underbara ting samt om nästan alla djur, som lefva i Norden och deras natur* (1555; Uppsala: Almqvist och Wiksell, 1909–51), pts. 1–5, 2:14; quotation on 14. Originally written in Latin, the book eventually reached twenty known editions (up to the year 1669) and was translated into French, Italian, Dutch, German, and English. See Oscar Almgren et al., "Utgifvarnas förord," ibid., pt. 1, iii.

55. Magnus, *Historia om de nordiska folken*, 14–15.

56. Agricola also made reference to this work in his *De re metallica*: "In some of our mines, however, though in very few, there are other pernicious pests. These are demons of ferocious aspect, about which I have spoken in my book De Animantibus Subterraneis." Georgius Agricola, *De re metallica: Translated from the first Latin edition of 1556 . . . by Herbert Clark Hoover . . . and Lou Henry Hoover* (1556; New York: Dover Publications, 1950), 217; quotation in Hoover's translation. *De animantibus subterraneis* was usually published together with *De re metallica* and was included already in its first edition of 1556. Except for a short excerpt, it was not included in the Hoover translation. H. C. Hoover and L. H. Hoover, introduction to *De re metallica*, ibid., vii. For the excerpt, see ibid., 217.

57. Georgius Agricola, "Die Lebewesen unter Tage," in G. Agricola, *Vermischte Schriften*, vol. 1, ed. Hans Prescher, trans. Georg Fraustadt (Berlin: VEB Deutcher Verlag der Wissenschaften,, 1961), 160–230. Original title: *De animantibus subterraneis liber* (1549). On the basilisk and dragon, 194–95. On Agricola's dependency on Aristotle, Rolf Hertel, "Zoologische Einführung," ibid., 141–54, esp. 141–43.

58. Agricola, "Die Lebewesen unter Tage," 199.

59. Ibid., 199–200.

60. Agricola used the expression that "their garment itself was made out of thicker cloth." Ibid., 199.

61. There were additional types of hybrid creatures. Paracelsus, "Tractatus IV. Von dem underscheit der corporum und spirituum," 350–51. Paracelsus, "Liber de nymphis, sylphis, pygmaeis et salamandris," 118, 124.

62. Paracelsus, "Liber de nymphis, sylphis, pygmaeis et salamandris," 120, 124–26.

63. Oswald Croll, *Basilica chymica: Oder Alchymistisch Königlich Kleÿnod: Ein Philosophisch durch sein selbst eigene erfahrung confirmierte und bestät-*

tigte Beschreibüng und gebrauch der aller fürtrefflichen Chimischen Artzneÿen so auß dem Liecht der Gnaden und Natur genommen in sich begreiffen (Frankfurt: Gottfried Tampachen, [1629?]), on 13.

64. Häll, *Skogsrået, näcken och djävulen*, 35, 252–56.

65. Bure, "Anteckningar af Johannes Thomae Agrivillensis Bureus," quotation on 15.

66. This was common practice for narrators of stories of this type. Häll, *Skogsrået, näcken och djävulen*, 48.

67. Bure held trolls to be demonic representatives of the wilderness and of the powers of chaos that ruled outside of the bounds of the Christian community. Karlsson, *Götisk kabbala*, 100–101, 128–29, 171, 239–43. On the differences between trolls and men in Scandinavian (Swedish/Finnish) folklore, see Bo Lönnqvist, "17. Troll och människor," in B. Lönnqvist, *De andra och det annorlunda: Etnologiska texter* (Helsingfors: Ekenäs tryckeri, 1996), 149–57.

68. In accounts gathered by folklore researchers in eastern Scandinavia in the nineteenth and twentieth centuries, the trolls were considered social beings, living alone or in family groups of various sizes, and leading lives similar to those of their human neighbors. Camilla Asplund Ingemark, *The genre of trolls: The case of a Finland-Swedish folk belief tradition* (Åbo: Åbo Akademi University Press, 2004), 7–10. For an overview of the literature, ibid.,13–21. For an overview of spirits associated with nature in Swedish nineteenth-century folklore, Sigurd Erixon, "Some examples of popular conceptions of sprites and other elementals in Sweden during the 19th century," in Å. Hultkrantz, ed., *The supernatural owners of nature: Nordic symposium on the religious conceptions of ruling spirits (genii loci, genii speciei) and allied concepts* (Stockholm: Almqvist and Wiksell, 1961), 34–37.

69. Bure, "Anteckningar af Johannes Thomae Agrivillensis Bureus." These annotations were made the same year as the previous one on the hurt troll, on 79.

70. Paracelsus, "Liber de nymphis, sylphis, pygmaeis et salamandris," 116–17, 128, 130–31, 134–36.

71. For a wealth of narratives about human encounters with keeper entities, see Häll, *Skogsrået, näcken och djävulen*, esp. 312–408; on clothing, 398–400. See also Carl-Herman Tillhagen, *Järnet och människorna: Verklighet och vidskepelse* (Stockholm: LTs förlag, 1981), 117–18, 128. Tillhagen identifies Olaus Magnus's trolls as the same type of entity as *gruvrådan* (mine keepers): ibid., 122, 125–26. According to Tillhagen, however, Swedish beliefs about the *gruvrådan* had grown out of the complex of beliefs surrounding *skogsrådan* (the forest keeper), hence the dissimilarities from German beliefs. On the wider context of keeper entities, see Åke Hultkrantz, ed., *The supernatural owners of nature: Nordic symposium on the religious conceptions of ruling spirits (genii loci, genii speciei) and allied concepts* (Stockholm: Almqvist och Wiksell, 1961). On the keeper of the Falun mine, see Daniels Sven Olsson, *Falun mine* (Falun: Stiftelsen Stora Kopparberget, 2010), 7–9.

72. Tillhagen, *Järnet och människorna*, 123, 134.

73. Ibid., 103–36.

74. Häll, *Skogsrået, näcken och djävulen*, 279–455. On the corporeality of these beings, see 394, 396, and throughout the volume.

75. In a 1570 case, countermagic was successfully used to lift a curse that had halted production at the silver mine of Sala. Such officially sanctioned use of magic is, however, rare in Swedish sources, Ankarloo, *Trolldomsprocesserna i Sverige*, 50.

76. Sven Rinman, *Bergwerks lexicon*, vol. 1 (Stockholm: Johan A. Carlbohm, 1788), 1002. *Svenska akademiens ordbok*, Internet ed., "nickel," http://g3.spraakdata.gu.se/saob/ (webpage visited 2005.08.30). Erich Odhelius to Urban Hiärne, Leipzig, 1 Oct. 1684, in Gjörwell, *Det swenska biblioteket* 2:263–69; on 269. *Svenska akademiens ordbok*, Internet ed., "kobolt," http://g3.spraakdata.gu.se/saob/ (webpage visited 2005.08.30).

77. Urban Hiärne, *Den korta anledningen / til Åthskillige malm och bergarters / mineraliers och jordeslags efterspörjande och angifwande / beswarad och förklarad / jämte deras natur / födelse och i jorden tilwerckande / samt uplösning och anatomie i giörligaste mått skrifwen* (Stockholm, 1702).

78. Ibid., 29, 120–22.

79. Ibid., 55.

80. Ibid.

81. Ibid., 56, 129–30.

82. Ibid., 63. In their immediate context, the Stockholm witch trials of 1675–76 were part of an attempt to control what was considered a larger outbreak of witchcraft, centered in particular around the northern part of Sweden (Norrland), the province of Dalarna, and Stockholm. They were the culmination of a juridical process in which thousands of people were heard before the courts, and at the end about 240 were executed. There were also processes occurring in other parts of Sweden. Figures exclude 38 people executed in Bohuslän, which traditionally was under Danish law. Ankarloo, *Trolldomsprocesserna i Sverige*, 23, 28, 228.

83. Some historians and many older popular books have presented Hiärne as the one who ended the witch trials in Stockholm, and even in Sweden at large. See, e.g., Bror Gadelius, *Urban Hjärne och häxprocesserna i Stockholm 1676* (Isaac Marcus: Stockholm, 1909), 1–39, esp. 1, 38. This view derives from the here quoted source and Hiärne's autobiography, although an account by Hiärne's descendant Daniel Tilas has also been quoted as the original source. Hiärne, "Salig Landzhöfding Urban Hiernes Lefwernesbeskrifning," 165–67. Daniel Tilas, "Anteckningar ur Hiärne Ättens Minne," in [Adlersparre], *Historiska samlingar*, 5:432–42 . Hiärne, however, did not participate during much of the proceedings, and his main contribution to the commission happened toward the end. It was instead the clergyman Noraeus and two other clergymen who led the commission to the position it eventually would take, i.e., the freeing of the accused witches. "Kongl. Commissorial Rättens Ransakning," 331–33. Ankarloo, *Trolldomsprocesserna i Sverige*, 204–11. Lindroth, "De stora häxprocesserna," 160–61.

84. Hiärne, "Kort betenckiande öffwer de anfechtade barnens i Stockholm," in [Adlersparre], *Historiska samlingar*, 5:409–32, esp 417, 429–30; quotation on 429.

85. Urban Hiärne, *Märkvärdigheter hos sjön Vettern*, trans. from Latin by Carl Stubbe (Stockholm: Norstedts, 1916). Originally published as "Memorabilia nonnulla lacus Vetteri," *Philosophical Transactions* 24, no. 298 (1704–5).

86. Magnus, *Historia om de nordiska folken.* On Vättern, see pt. 1, 61, 106, 165; pt. 4, 187–88.

87. Hiärne, *Märkvärdigheter hos sjön Vettern,* 11–14, 16.

88. Ibid., 14–15.

89. Ibid., 18; quotation on 25.

90. Ibid., 9.

91. Savin, "Gud i Naturens laboratorium," 118.

92. Principe, *Aspiring adept,* 41, 61–62, 206.

93. Tomas Mansikka, "'Helias cum veniet restitutet omnia': Paracelsism, alkemi och reform i 1600-talets protestantism" (Licenciate diss., Åbo akademi, 1998). Brian Copenhaver, "Natural magic, hermetism, and occultism in early modern science," in D. C. Lindberg and R. S. Westman, eds., *Reappraisals of the scientific revolution,* 261–301 (Cambridge: Cambridge University Press, 1990), 284–85.

94. Savin, *Fortunas klädnader,* 15, 17, 23, 36–38, 131, 167–207. David Dunér, *Världsmaskinen: Emanuel Swedenborgs naturfilosofi* (Lund: Nya Doxa, 2004), 186–87.

CHAPTER 3

1. William R. Newman, *Gehennical fire: The lives of George Starkey, an American alchemist in the scientific revolution* (Cambridge: Harvard University Press, 1994), 4–13. These transmutation histories were, in the words of Lawrence Principe, often "painstakingly precise, noting exact times, places, persons (often rank and station) in attendance, the quantity of gold or silver produced, and so forth. Other examples are public, having taken place at some court or assembly, and were sometimes commemorated by the striking of coins or medallions." Principe, *Aspiring adept,* 93. See also Pamela H. Smith, *The business of alchemy: Science and culture in the holy roman empire* (Princeton: Princeton University Press, 1994) 45.

2. Principe, *Aspiring adept,* 77–80.

3. Nummedal, *Alchemy and authority,* 33–34; quotation on 4. Smith, *Business of alchemy,* 6–9, 23. On the relationship between chymistry and medicine, Newman, *Gehennical fire,* xii.

4. On alchemy's popularity in early moder Europe, Newman, *Gehennical fire,* 3. See also Smith, *Business of alchemy.* For an interesting discussion of the sixteenth century, Bruce T. Moran, "German prince-practitioners: Aspects in the development of courtly science, technology, and procedures in the Renaissance," *Technology and Culture* 22, no. 2 (1981): 267, 268, 273.

5. Nummedal, *Alchemy and authority,* 9, and other places. Older research tended to describe interest in alchemy among territorial rulers as a form of escapism. Recent research has shown this picture wrong. On the contrary, it was often the rulers with the greatest interest in centralization, state building, and economic reform who also supported alchemy. Ibid., chap. 3.

6. Böttger is the protagonist of a widely read work of popular history, Janet Gleeson, *The Arcanum: The extraordinary true story* (New York: Warner,

1998). For an important corollary to Gleeson's narrative, see Lawrence W. Principe, "Transmuting history," *Isis* 98 (2007): 779–87.

7. The reorganization of state administrations in the seventeenth and eighteenth centuries is a vast topic. For a recent overview of the Swedish seventeenth century, see Mirkka Lappalainen, *Släkten — makten — staten: Creutzarna i Sverige och Finland under 1600-talet* trans. Ann-Christine Relander (Stockholm: Norstedts, 2007), 32–48, 137–40. For the eighteenth century, see Maria Cavallin, *I kungens och folkets tjänst: Synen på den svenske ämbetsmannen 1750–1780* (Göteborg: Historiska institutionen, 2003), 36–44. For the Holy Roman Empire, Andre Wakefield, *The disordered police state: German cameralism as science and practice* (Chicago: University of Chicago Press, 2009), chap. 1, esp. 16–17, 21.

8. For the period covered in this study, ca. 1680–1760, it can be noted that between 1680 and 1718 Sweden was an absolute monarchy, and the Council of the Realm was primarily an advisory body that the king could draw on, and that could govern the Realm in his absence. Although it was assumed that the councilors would have thorough knowledge of their specific field of competence, the Council was always dominated by the most influential aristocrats of the high nobility. In 1718–19 the constitution was changed and an era of parliamentarian rule began, which lasted until 1772. The king was required to select his councilors from a list provided by the Parliament (Riksdagen), and the Council of the Realm, not the king, became the country's real center of executive power, except when the Parliament was in session. Simultaneously the position of councilor was separated from that of president of a Bureau. There was still an overlap between the groups, but whereas the president of a Bureau still was expected to have a specialist's knowledge in his field, a councilor no longer had to have such competence. This opened up the Council of the Realm to full-time politicians. Ludvig Stavenow, *Om riksrådsvalen under frihetstiden: Bidrag till svenska riksrådets historia* (Uppsala: Almqvist och Wiksell, 1890), 1–5, 12–17, 22. There was one exception: the head of the Royal Chancellery continued to have a permanent seat in the Council of the Realm. Ibid., 22.

9. Sven Gerentz, *Kommerskollegium och näringslivet: Minnesskrift utarbetad av Sven Gerentz på uppdrag av Kungl. Kommerskollegium till erinran om Kollegii 300-åriga ämbetsförvaltning 1651–1951* (Stockholm: Kommerskollegium, 1951), 15, 18–19. Thomas Kaiserfeld, *Krigets salt: Salpetersjudning som politik och vetenskap i den svenska skattemilitära staten under frihetstid och gustaviansk tid* (Lund: Sekel, 2009), 18–26.

10. Christoph Bartels, "The production of silver, copper, and lead in the Harz Mountains from late medieval times to the onset of industrialization," in Klein and Spary, *Materials and expertise in early modern Europe*, 78–79; quotation on 79. See also Bartels, *Vom früneuzeitliche Montangewerbe zur Bergbauindustrie*, 446–47. Wolfhard Weber, *Innovationen im frühindustriellen deutschen Bergbau und Hüttenwesen: Friedrich Anton von Heynitz Studien zu Naturwissenschaft, Technik und Wirtschaft im Neunzehnten Jahrhundert Herausgegeben von Wilhelm Treue 6* (Göttingen: Vanderhoeck und Ruprecht, 1976), 107.

11. Bartels, *Vom früneuzeitliche Montangewerbe zur Bergbauindustrie*, 55, 79. Hans Baumgärtel, *Bergbau und Absolutismus: Der sächsische Bergbau in der zweiten Hälfte des 18. Jahrhunderts und Massnamen zu seiner Verbesserung nach dem Siebenjährigen Kriege* Freiberger Forschungshefte D 44 Kultur on Technik (Leipzig: VEB Deutscher Verlag für Grundstoffindustrie, 1963), 25–26.

12. Baumgärtel, *Bergbau und Absolutismus*, 16–17.

13. Lindroth, "Urban Hiärne och Laboratorium Chymicum," 56–60.

14. On the manufacture of medicines as the main task of the laboratory, see ibid., 57.

15. Ibid., 54–55. Hiärne, as quoted through Lindroth, 54.

16. Nummedal, *Alchemy and authority*, chap. 5, contains an overview of such laboratories.

17. Hiärne had become assessor extraordinary already in 1675. J. A. Almquist claims that the Bureau had already planned to make him the leader of the laboratory. Lindroth, on the contrary, is of the opinion that there was no such plan, and that Councilor of the Realm Edmund Gripenhielm had appointed Hiärne despite the misgivings of the Bureau, which had doubted that Hiärne was up to the job. Lindroth, "Urban Hiärne och Laboratorium Chymicum," 55, 61–65.

18. Ibid., 61–64, 68–69, 80–85. This impressive laboratory was operational until 1708, when Hiärne decided to relocate the laboratory to a building closer to his home. The palace was sold in 1712, but Hiärne continued to do laboratory work until at least 1720 (he was by then almost eighty). Ibid., 111–13.

19. Lindroth discusses Sybellista's gold-making plans, 1665–68; there are also general discussions at the Bureau concerning alchemy, which took place between Fleming and Bernhard Below in 1670, when Fleming was president of the Bureau. Ibid., 58–60.

20. Smith, *Business of alchemy*, 6–9, 23; quotation on 7.

21. Ibid., 199. Nummedal, *Alchemy and authority*, chap. 6. Wakefield, *Disordered police state*, 5–8.

22. Wakefield, *Disordered police state*, 8.

23. Smith, *Business of alchemy*, 6–9, 23.

24. Erich Odhelius was the third son of a father of the same name: Ericus Odhelius, professor of theology and orientalist at Uppsala. His father died early (1666), and his mother, Margareta Laurelia, remarried. Her second husband, Nicolaus Rudbeckius, would later become bishop of Västerås and was also a brother of Olof Rudbeck the elder. Odhelius was ennobled in 1698 and took the name Odelstierna from that year. Nevertheless I call him Odhelius throughout the book because to do otherwise would confuse the analysis and descriptions of social interactions between him and others if his name conveyed the impression that he was a nobleman in his youth. Indeed his Latinized name served as a marker to early modern Europeans that he was a scholar or came from a family of scholars. I have, however, followed convention with other actors, e.g., Swedenborg, Polhem, and von Bromell.

25. Elsa-Britta Grage, "Odelstierna, Erich," in *Svenskt biografiskt lexikon*, 28:34–37. Mats Ola Cajdert, "En Hiärne-lärjunge i Europa: Erik Odelstiernas Brev till Urban Hiärne 1683–1687," *Personhistorisk tidskrift* 92 (1996): 1–6.

Although the correspondence discussed below is one-sided (i.e., only Odhelius's letters remain), Hiärne's strong regard for and interest in Odhelius is evident from his diaries. See, e.g., Hiärne's diary of October 1688, in which Odhelius is mentioned at the beginning of almost every new entry. Urban Hiärnes manuskript D 14, UUB.

26. Erich Odhelius to Urban Hiärne, 26 Dec. 1683, in Gjörwell, *Det swenska biblioteket*,1:314.

27. Odhelius to Hiärne, 9 Jan.1684, ibid., 1:316–17. On new books, Odhelius to Hiärne, 4 Jun. 1684, ibid.,257. On forwarding texts, Odhelius to Hiärne, 22 Mar. 1684, ibid., 323. On recommending assistants, Odhelius to Hiärne, 20 Feb. 1684, ibid., 320.

28. Odhelius to Hiärne, 26 Dec. 1683, ibid., 1: 313.

29. X273 "Bergmästaren Erich Odhelii berättelse till Kongl. Maj:t om en resa till Tyskland, Ungern, Tirol, Biscaien, Engelland och Holland för vinnande af kännedom om Bergvärkens och metallernas der i provincierna tillstånd och afförsel," KB.

30. Examples are Johan Rantze's sketches of mines in Saxony, the Hartz, and Hungary dating from 1691. Thomas Cletcher wrote on European mines in 1696, Jacob Hertzen of his German travels in 1697, and Edmund Gripenhielm made a report on his travels in Hartz, Mansfeld, Hesse, Saxony, and Bohemia. Martin Fritz, introduction to M. Fritz, ed., *Iron and steel on the European market in the seventeenth century: A contemporary Swedish account of production forms and marketing* (Stockholm: Berlings, 1982), 9–13.

31. Baumgärtel, *Bergbau und Absolutismus*, 34–35, 49, 78–79, 96. For an overview of Freiberg in the decades around 1700, Walther Herrmann, *Bergrat Henckel: Ein wegbereiter der Bergakademie* Freiberger Forschungshefte D 37 Kultur on Technik (Berlin: Akademie-Verlag, 1962), 11–32.

32. Baumgärtel, *Bergbau und Absolutismus*, 10–15.

33. Letters of recommendation from Hiärne ensured that he was well received by several learned men on the way, e.g., in Copenhagen, Hamburg, and Lübeck. Odhelius to Hiärne, 13 Oct. 1683, in Gjörwell, *Det swenska biblioteket*, 1:299–307.

34. Odhelius to Hiärne, 11 Apr. 1685, ibid., 2:281–85. Linck seems to have run a large business. In a letter, Odhelius mentioned that during an epidemic of dysentery, the pharmacy prepared more than a hundred recipes each day. Odhelius to Hiärne, 23 Aug. 1684, ibid., 263–65.

35. Odhelius to Hiärne, Leipzig, 1 Nov. 1683, 26 Dec. 1683, 20 Feb. 1684, 1:307–15, 317–20, 319. See also Odhelius to Hiärne, 25 Jun. 1684, ibid., 2:259–263.

36. Odhelius to Hiärne, Leipzig, 22 Mar. 1684, ibid., 1:322–23. The mineral collection is also discussed in other letters, see, e.g., Odhelius to Hiärne, Leipzig, 7 May 1684, ibid., 1:325–26. Odhelius to Hiärne, Leipzig, 7 May 1684, ibid., 2:330.

37. Odhelius to Hiärne, Leipzig, 7 May 1684, ibid., 1:326. Odhelius's description of the knowledge of the apothecary bears similarities to that of historian Harold J. Cook, *Matters of exchange: Commerce, medicine, and science in the Dutch golden age* (New Haven: Yale University Press, 2007), 31.

38. On the role of assaying in the early modern smelting process, Bartels, "Production of silver, copper, and lead," 87–89. Baumgärtel, *Bergbau und Absolutismus*, 34–35. For a course in assaying held at the Bureau of Mines in the first half of the eighteenth century, X277 Prober-konsten sammanskrefwen af Jacob Fischer proberare uti Kongl. Bärgs Collegio., KB.

39. Odhelius to Hiärne, Leipzig, 7 May 1684, in Gjörwell, *Det swenska biblioteket*, 1:325–31, esp. 326.

40. Odhelius to Hiärne, 26 Dec. 1683, 9 Jan. 1684, 20 Feb. 1684, 22 Mar. 1684, 7 May 1684, ibid.,1:313, 317, 317–20, 324, 328, respectively.

41. Odhelius to Hiärne, Leipzig, 23 Aug. 1684, ibid., 2:264.

42. Odhelius to Hiärne, 11 Oct. 1684, 17 Dec. 1684, ibid., 2:269–70, 275. He continued to fret about going to Freiberg, and asked Hiärne about information on the Swedish minerals in case he would be asked questions about them. He also asked Hiärne to send him a box of Swedish minerals and a letter of recommendation to one Süssmilch, a man he had chosen as the most suitable teacher. Odhelius to Hiärne, 17 Dec. 1684, 11 Apr. 1685, ibid.,2:276, 281–85.

43. Odhelius to Hiärne, Leipzig, 1 Oct. 1684, ibid., 2:269.

44. Bartels, *Vom früneuzeitliche Montangewerbe zur Bergbauindustrie*, 52, 279–81.

45. Herrmann, *Bergrat Henckel*, 24.

46. Baumgärtel, *Bergbau und Absolutismus*, 78–79.

47. Odhelius to Hiärne, 21 May 1685, 5 Dec. 1685, ibid., 2:285–87, 293–94, respectively.

48. There was, however, a less formalized structure in existence previously; the oldest remaining request for such a stipend is from 1681. Baumgärtel, *Bergbau und Absolutismus*, 79. Weber, *Innovationen im frühindustriellen deutschen Bergbau und Hüttenwesen*, 155.

49. A number of Swedish travelers visited Freiberg in the decades following Odhelius's first visit. In one of his books Hiärne narrates a conversation between Swedish traveler Samuel Buschenfelt and von Schönberg. Urban Hiärne, *Den besvarade och förklarade anledningens andra flock: Om jorden och landskap i gemeen* (Stockholm: Mich. Laurelio, 1706), 352–53.

50. Jacob Orrje, *A mechanical state: The mathematical sciences and public office in the eighteenth-century Sweden* (forthcoming). Other Bureaus also awarded grants for foreign travel; Gerentz, *Kommerskollegium*, 66.

51. Weber, *Innovationen im frühindustriellen deutschen Bergbau und Hüttenwesen*, 155.

52. Odhelius to Hiärne, 27 Sep.1685, 3 Feb. 1686, 5 Dec. 1685, in Gjörwell, *Det swenska biblioteket*, 2:288–90, 293–94; quotation on 2:295–96.

53. It is difficult to ascertain from the letters what Linck's experiments were about, and what aims he could have had in conducting them. Odhelius to Hiärne, 26 Dec. 1683, 22 Mar. 1684, ibid., 1:312, 320–25.

54. Odhelius to Hiärne, 23 Jan. 1685, 11 Apr. 1685, ibid., 2:279, 281–85.

55. It is written, on the copy in Boyle's hand, that he had received it from the author "himselfe." Principe, *Aspiring adept*, 112. I thank Lawrence Principe for pointing this out to me.

56. Odhelius to Hiärne, 13 Oct. 1683, 26 Dec. 1683, 22 Mar. 1684, in Gjörwell, *Det swenska biblioteket*, 1:306, 312, 320–25. The term *Pseudo-Adepti* is used on 322.

57. Already pseudo-Geber's thirteenth-century *Summa perfectionis* emphasized that "the philosophers' stone is a donum dei—a gift from God—which he 'extends to and withdraws from who He wishes.'" Newman, *Gehennical fire*, 3. See also Principe, *Aspiring adept*, 108. On Odhelius's view, Odhelius to Hiärne, 11 Nov. 1685, in Gjörwell, *Det swenska biblioteket*, 2:293. Odhelius to Hiärne, Leipzig, 7 May. 1684, ibid.,1:327.

58. Odhelius to Hiärne, 26 Dec. 1683, 20 Feb. 1684, in Gjörwell, *Det swenska biblioteket*,1:313, 318. Odhelius to Hiärne, 25 Jun 1684, ibid., 2:262.

59. Odhelius to Hiärne, 22 Mar. 1684, ibid., 1:322.

60. The man's name is given as Stanziani. Odhelius to Hiärne, 23 Jan. 1685, ibid., 2:280.

61. Odhelius to Hiärne, 11 Nov. 1685, ibid., 22:292–93.

62. Starcke may possibly be the builder and mechanic Johann Georg Starke (1650–95); *Sächsische biographie*, http://saebi.isgv.de (accessed 11 Nov. 2013). Odhelius to Hiärne, Freiberg, 11 Nov. 1685, in Gjörwell, *Det swenska biblioteket*, 2:292–93; quotation "Although" on 292; quotation "Time will tell" on 293.

63. Odhelius to Hiärne, Freiberg, 11 Nov. 1685, in Gjörwell, *Det swenska biblioteket*, 2:293.

64. Joh. Joachim Becher, *Närrische Weißheit Und Weise Narrheit: Oder Ein Hundert so Politische als Physicalische/ Mechanische und Mercantilische Concepten und Propositionen/ deren etliche gut gethan/ etliche zu nichts worden* (1707), 144, 172; quotation "öffentliche Betrüger und Sophisten" on 172. In the paragraph which Odhelius refers to in Becher's text, the man's name is given as Marsali. Odhelius to Hiärne, 11 Nov. 1685, in Gjörwell, *Det swenska biblioteket*, 2:292–93; quotations on 292, 293. Becher is the main protagonist of Smith, *Business of alchemy*; this book is discussed on 260–61.

65. That Odhelius shared many of his patrons' view of Paracelsus is evident from several passages in the letters, e.g., when Odhelius showed irritation at Helmont's denial of Paracelsus's Tartarum and said that he believed Helmont has not approached the heights to which Paracelsus had risen. Odhelius to Hiärne, 1 Oct. 1684, in Gjörwell, *Det swenska biblioteket*, 2:268. Odhelius also credited Paracelsians with higher knowledge than other medical practitioners. Ibid. Paracelsus authority was also a natural point of reference for Odhelius, who discussed with Hiärne whether later authors were in agreement with Paracelsus or not. Odhelius to Hiärne, 22 Mar. 1684, in Gjörwell, *Det swenska biblioteket*, 1: 324.

66. Hiärne, *Defensionis paracelsicae prodromus*. Hiärne's text was a response to a text written by the provincial physician Magnus Gabriel von Block, and it was written in a conciliatory tone. Hiärne claimed that he agreed with what he took to be Block's main objective, the refutation of contemporary fortune-tellers. He also stated that most of Block's arguments against Paracelsus were taken from the works of Hermann Conring and Gabriel Naudé, and that therefore it would be more suitable to refute these authors. Hiärne, *Defensio-*

nis paracelsicae prodromus, 1–2. Hiärne's initial response to Block's pamphlet, however, had been furious, and it is possible that he continued to work against Block for some years after. See Sten Lindroth, "Hiärne, Block och Paracelsus: En redogörelse för paracelsusstriden 1708–1709," *Lychnos* (1941): 211–23, 225–29.

67. Hiärne, *Defensionis paracelsicae prodromus*, 3.

68. Ibid., 4–5.

69. Hiärne's list of opponents included Hermann Conring, Gabriel Naudè, and Andreas Libavius. Ibid., 2, 4.

70. "Fratrem, sororem & poculum amoris in montibus Chymicorum, foetum, Draconem, ejus caudam & venenum, nigrum nigrius nigro &c." Odhelius to Hiärne, Leipzig, 7 May 1684, in Gjörwell, *Det swenska biblioteket*, 1: 327.

71. Odhelius to Hiärne, 7 May 1684, ibid., 1:327.

72. Ibid., 1:325–31. Odhelius to Hiärne, Leipzig, 25 Jun 1684, ibid., 2:261.

73. Odhelius's views show similarities to those of Robert Boyle in this regard; Principe, *Aspiring adept*, 46–47.

74. The above passage has been interpreted for clarity. Odhelius does not speak of a seminal force that creates gold but simply of a force. Odhelius to Hiärne, 21 May 1685, in Gjörwell, *Det swenska biblioteket*, 2:285–87.

75. Jean d'Espagnet, *Enchyridion physicae restitutae; or, The summary of physicks recovered. Wherein the true harmonie of nature is explained, and many errours of the ancient philosophers, by canons and certain demonstrations, are clearly evidenced and evinced. / Written in Latine by the Duke of Espernon; and translated into English by Dr. Everard* (London: W. Bentley, [1651]).

76. Odhelius to Hiärne, 7 May 1684, in Gjörwell, *Det swenska biblioteket*, 1:327. In fact, Odhelius's position seems to contradict that of Paracelsus. Compare Paracelsus in the work *Von den natürlichen Dingen* on the difference between transmuting iron into copper, and iron into gold: "Many arts are withheld from us because we have not ingratiated ourselves to God so that He would make them manifest to us. To make iron into copper is not as much as to make it into gold. Hence what is less God has allowed to emerge." As quoted in Newman, *Gehennical fire*, 4.

77. Odhelius to Hiärne, 20 July 1690, in Gjörwell, *Det swenska biblioteket*, 2:312–13.

78. "Protestant belief did not hold that the sacred did not intrude into the secular world, simply that it did not do so at human behest and could not automatically be commanded." R. W. Scribner, "The reformation, popular magic, and the 'disenchantment of the world,'" in R. W. Scribner, *Religion and culture in Germany (1400–1800) by R. W. Scribner edited by Lyndal Roper* (Leiden: Brill, 2001) 346–65: 354.

79. Odhelius to Hiärne, 7 May 1684, in Gjörwell, *Det swenska biblioteket* 1:326.

80. Lindroth, "Urban Hiärne och Laboratorium Chymicum," 98–101, 104–6.

81. Urban Hiärne, *Actorum Laboratorii Stockholmiensis Parasceve. Eller Förberedelse Til de undersökningar som uthi Kongl. Laboratorio äro genomgångne samt underrättelse om de förnämbste Grundstycken uthi Chymien*

(Stockholm: Michaele Laurelio, 1706), 11. See also Paracelsus, "Das Buch De Mineralibus," in *Theophrast von Hohenheim gen. Paracelsus, Sämtliche Werke* *1* , pt. 3, 32–33.

82. Principe, *Aspiring adept*, 58, 61.

83. Hiärne, *Actorum Laboratorii Stockholmiensis Parasceve*, 42–43. On Becher, ibid., 47.

84. Copenhaver, "Natural magic, hermetism, and occultism in early modern science," 269–71.

85. Odhelius to Hiärne, 11 Nov. 1685, in Gjörwell, *Det swenska biblioteket*, 2:293.

86. Compare discourse on cameralists in Wakefield, *Disordered police state*, 8, 12.

87. Sten Bielke and his successor Claes Fleming both supported Hiärne. Wrede was Fleming's successor. Letter from Hiärne to Olof Hermelin, 24 Apr. 1707, published in *Historisk Tidskrift* (1882): 265.

88. Lindroth, *Gruvbrytning och kopparhantering vid Stora Kopparberget intill 1800-talets börja: del 1 Gruvan och gruvbrytningen* (Uppsala: Almqvist och Wiksell, 1955), 166–68. For a recent overview of Bureau of Mines efforts at the mine in the eighteenth century, see Hjalmar Fors, "'Away, away to Falun!' J. G. Gahn and the application of Enlightenment chemistry to smelting," *Technology and Culture* (2009): 549–68.

89. Lindroth, *Gruvbrytning och kopparhantering*, 40–50.

90. Newman, *Gehennical fire*, 4.

91. Hiärne discusses the traditional Falun method in Hiärne, *En ganska liten bergslykta: förmedelst hvilken man sig uti den mörka bergsverkshandtering själf leda och lysa kan*, ed. Carl Sahlin (Örebro: Örebro Dagblads tryckeri, 1909), 32–33.

92. Ibid., 33. See also Nils Zenzén, "Från den tid, då vi skulle transmutera järn till koppar och få lika mycket silver i Sverige som gråberg," *Med Hammare och Fackla* 7 (1936): 91–92.

93. Zenzén, "Från den tid, då vi skulle transmutera järn till koppar," 92–98.

94. Odhelius to Hiärne, 7 May 1690, in Gjörwell, *Det swenska biblioteket*, 2:308–10. Zenzén, "Från den tid, då vi skulle transmutera järn till koppar," 102–3.

95. X273 Bergmästaren Erich Odhelii berättelse, 20–23.

96. Ibid., 3–4, 20.

97. Odhelius to Hiärne, 7 May 1690, 20 Aug. 1690, in Gjörwell, *Det swenska biblioteket*, 2:308–10, 314–15. See also a previous letter in which Odhelius stated that the colonel's work was of little or no interest to the Bureau of Mines. Odhelius to Hiärne, 20 July 1690, ibid., 2:312–13.

98. Odhelius to Hiärne, 20 Aug. 1690, ibid., 2:314–15; quotation on 315.

99. "Kongl. Maj:ts och Riksens Bergs-Collegii Relation om Bergs wäsendets tilstånd i Riket under Konung Carl XI:tes tid, dat. den 30 Sept. 1697," in [Samuel Loenbom], ed., *Handlingar til Konung Carl XI:tes Historia*, pt. 12 (Stockholm: Lars Salvius, 1772), 102–7.

100. Kunckel, a prominent chymist of European reputation, had been invited primarily because of his knowledge of smelting (although his abilities as

a chrysopoeian also were discussed at the Bureau). Kunckel had been made a mine councilor in 1692, and hence stepped in—at least formally—into the Bureau's hierarchy at the highest level. Simultaneously, Kunckel's son and their associate Conrad Müller were also employed and given lower, albeit impressive, titles. None of the men received a salary, merely the promise that they would be compensated for their expenses if found deserving. In 1693 Kunckel was also ennobled under the name of Kunckel von Löwenstern. It was mostly Müller who interacted with the Bureau, and it is unclear whether Kunckel began any work in Sweden prior to December 1695, when he became engaged in the improvement of the copper process at Falun. He left the country in October 1696. Zenzén, "Från den tid, då vi skulle transmutera järn till koppar," 110–13, 116, 149–51.

101. Odhelius to Hiärne, 13 Oct. 1683, in Gjörwell, Det swenska biblioteket, 1:306.

102. Odhelius to Hiärne, 23 Sep. 1692, ibid., 2:327–28.

103. Letter from Hiärne to Olof Hermelin, 24 Apr. 1707, published in Historisk Tidskrift (1882): 265. On court intrigues against Hiärne in 1689, see Odhelius to Hiärne, Stockholm, 30 Oct. 1689, in Gjörwell, Det swenska biblioteket, 2:306–7. According to Lindroth, Wrangel supported alchemy but attempted to replace Hiärne with a client of his own. Lindroth, "Urban Hiärne och Laboratorium Chymicum," 87.

104. Kongl. Maj:ts och Riksens Bergs-Collegii Relation,"in [Loenbom], Handlingar, pt. 12, 149.

105. "Fortsättning till Slut, af Kongl. Maj:ts och Riksens Bergs-Collegii Relation om Bergs wäsendets tilstånd i Riket under Konung Carl XI:tes tid," in [Loenbom], Handlingar, pt. 13 (Stockholm: Lars Salvius, 1772), 109–10; quotation on 110.

106. Kongl. Maj:ts och Riksens Bergs-Collegii Relation" in [Loenbom], Handlingar, pt. 12, 141–43; quotation on 141. For Hiärne's report of this journey, see Urban Hiärne, "En kort Beskrifning af min Resa från Stockholm genom Uppland, Gestricke- och Helsingeland til Herjedalen, Norige och Jemtland, dädan til Medelpad och Ångermanland, sedan genom Helsingeland samma vägen hem igen" (1685), in Carl Gjörwell, Nya svenska biblioteket, vol. 1, pt. 2 (Stockholm: Peter Hessleberg, 1762), 233–59, 267–76.

107. "Fortsättning till Slut, af Kongl. Maj:ts och Riksens Bergs-Collegii Relation," in [Loenbom], Handlingar, pt. 13, 60. See also the narrative of how Hiärne was sent out to inspect innovative designs in smelting at the Sala silver works. "Kongl. Maj:ts och Riksens Bergs-Collegii Relation," in [Loenbom], Handlingar, pt. 12, 27–37, esp. 27.

108. "Kongl. Maj:ts och Riksens Bergs-Collegii Relation," in [Loenbom], Handlingar, pt. 12, 102–7.

109. Bergskollegiums arkiv, huvudarkivet, protocols for 1697, A1:35 RA, pp. 464–65, 1130–31. For the quoted discussion (of 23 Sep. 1697), see 1079–82.

110. The original text used the phrase "the nobility of the science" (Sciencens ädelhet). "Capable persons" is a translation of skickelige Personer. The term wettenskap, which corresponds to the German Wissenschaft, was also

used. "Fortsättning till Slut, af Kongl. Maj:ts och Riksens Bergs-Collegii Relation," in [Loenbom], *Handlingar*, pt. 13, 105.

111. This was to become a consistent policy. Lindqvist, *Technology on trial*, 118–34, esp. 121. Samuel Bring, "Bidrag till Christopher Polhems lefnadsteckning," in S. Bring, ed., *Christopher Polhem: Minnesskrift utgifven af Svenska Teknologiföreningen*, 3–119 (Stockholm: Centraltryckeriet, 1911), 25–26. Other Bureaus also awarded grants for foreign travel: Gerentz, *Kommerskollegium*, 66.

112. "Fortsättning till Slut, af Kongl. Maj:ts och Riksens Bergs-Collegii Relation," in [Loenbom], *Handlingar*, pt. 13, 105–6.

113. Nils Zenzén, "Studier i och rörande Bergskollegii Mineralsamling," *Arkiv för kemi, mineralogi och geologi* 8, no. 1 (1920): 12–15.

114. Hiärne, *Actorum Laboratorii Stockholmiensis Parasceve*, 1–5.

115. Ibid., 8–10.

116. Hiärne to Olof Hermelin, 24 Apr. 1707, published in *Historisk Tidskrift* (1882): 264–68; quotation on 267.

117. Björn Asker, "Paykull, Otto Arnold," in *Svenskt biografiskt lexikon*, 28:768–70. Lindroth, "Hiärne, Block och Paracelsus," 192–95. Hiärne's treatise was published in 1757 by Gjörwell; there are nine copies in the Uppsala University library.

118. Samuel Bark to Olof Hermelin, 11 May 1707, in Carl von Rosen, ed., *Bref från Samuel Bark till Olof Hermelin 1702–1708: Senare delen 1705–1708* (Stockholm: Norstedt, 1915), 185.

119. Hiärne to Olof Hermelin, 22 June 1707, published in *Historisk Tidskrift* (1882): 270–73. Hans Gillingstam, "Hermelin, släkt," in *Svenskt biografiskt lexikon*, 18:702.

120. Block had studied in Uppsala but made his career in Italy, where he claimed he had defended more than ten theses in Italian and Latin, at the academy in Rome that Ciampini had created under the protection of Queen Christina. After journeys to England and the Netherlands, and taking a medical degree in the University of Harderwijk, he returned to Sweden in 1702, where he gained a position at the health spa of Medevi. His chymical interests had been founded during his stay in Italy. Lindroth, "Hiärne, Block och Paracelsus," 198–201.

121. Hiärne also published a response to Block, *Defensionis Paracelciae Prodromus* (1709). This text was discussed in chapter 2. According to Lindroth, the German translation of Block's work was published in the town of Stade (1711). The mention in *Journal des scavans* was of February 1709, p. 361. Lindroth, "Hiärne, Block och Paracelsus," 196–98, 198n6, 202–4, 211–23, 225–29.

122. Jesper Svedberg, "First letter of Bishop Swedberg to Johan Rosenadler," dated 28 Feb. 1718, in R. L. Tafel, ed., *Documents concerning the life and character of Emanuel Swedenborg: Collected, translated, and annotated by R. L. Tafel* (Swedenborg Society, 1875), 1:155–59. On Swedenborg's and Hiärne's controversy, see also Jesper Svedberg, "Fifth letter of Bishop Swedberg to Johan Rosenadler," dated 2 Jan. 1720, ibid.,164–66.

123. He studied surgery and anatomy with G. Bidloo, botany and therapy with P. Hotton, and botany with P. Hermann. He went on hospital rounds

with F. Dekker, and learnt chymistry and pharmacy with J. Lemort. O. T. Hult, "Bromell, Magnus von," in *Svenskt biografiskt lexikon*, 6:393–94.

124. Bromell's first thesis was titled *De phlyctænis* under praeses G. Bidloo, 1700. His second thesis, *De non existentia spirituum* (praeses unknown), was defended the same year. Bromell matriculated as a student of medicine on 28 Apr. 1698. See "Volumina Inscriptorum" [shelf mark ASF 13], Universiteitsbibliotheek Leiden. The dissertations does not seem to have been preserved in the collections of Leiden's university library. I thank John A. N. Frankhuizen for obtaining this information for me.

125. Govard Bidloo, the praeses of Bromell's first dissertation, was professor of anatomy in Leiden and one of Europe's most renowned anatomists. Dániel Margócsy, "A museum of wonders or a cemetery of corpses: The commercial exchange of anatomical collections in early modern Netherlands," in Sven Dupré and Christoph Lüthy, eds., *Silent messengers: The circulation of material objects of knowledge in the early modern Low Countries* (Berlin: LIT Verlag, 2011), 185–215. On Bidloo's background and career, ibid., 185, 191–92.

126. Magnus von Bromell, *Lithographiae svecanae: Specimen primum, Calculos humanos, variaqve animalium concreta lapidea exhibens, juxta seriem atque ordinem, quo in Musaeo Metallico Bromeliano servantur* (Uppsala, 1726); M. Bromell, *Lithographiae svecanae: Specimen secundum, Telluris svecanae petrifacata lapidesque figuratos varios exhibens, juxta seriem atque ordinem, quo in Museo Metallico Bromeliano servantur* (Uppsala, 1727).

127. Bromell's family was wealthy, and after the death of his father and brother he became a man of independent means. Hult, "Bromell, Magnus von," 394–95. On his mineral collection, Zenzén, "Studier i och rörande Bergskollegii Mineralsamling," 13–14.

128. After Hiärne's death Bromell was officially appointed physician of the royal family (*Archiater*), preses of the Collegium Medicum, head of the Laboratorium Chymicum, and director of the mines of the province of Lapland. Hult, "Bromell, Magnus von," 392, 397. Concerning the value and cost of anatomical preparations during this time, see Margócsy, "Museum of wonders," 204–10.

129. Lindroth, "Urban Hiärne och Laboratorium Chymicum," 91.

130. D 1433 UUB [Anton von Swab, copy in the handwriting of Lars Schultze], Kort betänkande huru en yngling som täncker söka sin fortkomst vid Bergsväsende bör sin tid anlägga, 2, 5.

131. Manuscript, "Om Bergverken i Tyskland," Bergskollegiums arkiv huvudarkivet, D 6:9. (Daniel Tilas's collection) RA. The volume is part of a longer series consisting of relations from various countries in Europe. Germany was covered in four large volumes, D 6:9–D 6:13.

132. J. J. Ferber to P. W. Wargentin, 8 Dec. 1776, 12 Oct. 1780, 29. Mar. 1781, 25 Oct. 1781, 27 Apr. 1783, Wargentins arkiv, brev till P. Wargentin, E1:7, KVA.

133. An undated but very interesting instruction specified that young men who had their mind set on a career in mining should visit Hannover, Sachsen, Böhmen, and England, if given the opportunity. D 1433 UUB [Anton von

Swab, copy in the handwriting of Lars Schultze] Kort betänkande huru en yngling som täncker söka sin fortkomst vid Bergsväsende bör sin tid anlägga, 6. 134. Principe, *Aspiring adept*, 112–13, 134–36, 151, 163–64.

CHAPTER 4

1. Mechanical philosophy still held sway in Sweden during the decades around the middle of the eighteenth century. Rydén, "Enlightenment in practice," 11–14. As we will see in chapter 5, it also formed a productive framework for natural philosophical reinterpretations of nature.

2. For overviews, see e.g., Carolyn Merchant, *The death of nature: Women, ecology and the scientific revolution* (San Francisco: Harper Row, 1980), esp. chap. 9, and epilogue. Steven Shapin, *The scientific revolution* (Chicago: University of Chicago Press, 1996). Natural history, however, would not abandon the marvelous for the useful until some decades later; see, Benjamin Schmidt, "Inventing exoticism: The project of Dutch geography and the marketing of the world, circa 1700," in Smith and Findlen, *Merchants and marvels*, 357.

3. In classical Latin, the word *ingenium* signifies an innate character, natural talent, or even genius that often is associated to the ability to invent or create. During the early modern period, it could also be used to represent the creative power of the spirit. In the seventeenth century, the word *ingeniosus* had acquired a military connotation, signifying someone knowledgeable in the construction and handling of siege machinery. It was this military connotation that distinguished the *ingeniosus* from the more generalist *mechanicus*. It is also the root of the modern term *engineer* (in German and Swedish via the French rendering of the term, *ingénieur*). Here, *ingenious* is used as an adjective in its original Latin sense, but it also approaches its modern English sense of "clever." When discussing persons engaged in, e.g., engineering, the terms *mechanici* or *mechanical projectors* are instead used. Hélène Vérin, *La gloire des ingénieurs: L'intelligence technique du XVIe au XVIII siècle* (Paris: Albin Michel, 1993), 19–21. For the terms *engineer* and *mechanicus* in the Swedish context, Lindqvist, *Technology on trial*, 15.

4. Odhelius to Hiärne, 15 Nov. 1686, in Gjörwell, *Det swenska biblioteket*, 2:300–301; quotation on 300.

5. Odhelius to Hiärne, 27 Dec. 1686, ibid., 2:301–3; quotation on 302.

6. Steven Shapin, "The invisible technician," *American Scientist* 77 (1989): 560–63.

7. Odhelius to Hiärne, 27 Dec. 1686, in Gjörwell, *Det swenska biblioteket*, 2:301–3. A different interpretation of Odhelius's impression can be found in Rydberg, *Svenska studieresor till England*, 140–41.

8. X273 Bergmästaren Erich Odhelii berättelse, 557.

9. Odhelius to Hiärne, 27 Dec. 1686, 5 Dec. 1691, 2 Feb. 1691, in Gjörwell, *Det swenska biblioteket*, 2:301–3, 321–22, 322–24; quotation on 302. Rydberg, *Svenska studieresor till England*, 145n5.

10. Shapin and Schaffer, *Leviathan and the air-pump*. On meeting Hooke, see Odhelius to Hiärne, London, 23 Feb. 1692, in Gjörwell, *Det swenska bib-*

lioteket, 1:331–35. On Hooke, see ibid., and X273 Bergmästaren Erich Odhelii berättelse, 491.

11. X273 Bergmästaren Erich Odhelii berättelse, 559–60.

12. Dym characterizes the Royal Society's interest in mining and mineralogy as ambitious but highly dependent on Continental developments. Dym, *Divining science*, 140–41.

13. Odhelius to Hiärne, 23 Feb. 1692, in Gjörwell, *Det swenska biblioteket*, 1:331–35; quotation on 314.

14. Ibid.

15. Odhelius to Hiärne, 6 Jun. 1692, ibid., 1:336–39; quotation on 336.

16. Ibid.; quotation on 337.

17. Ibid., 1:336–39. As shown by this exchange, Hiärne could not have known about Homberg prior to receiving this letter. The claim presented in some biographies of Homberg—that he had studied with Hiärne in his Stockholm laboratory—must therefore be false.

18. Ibid. Quotation on 339. X273 Bergmästaren Erich Odhelii berättelse, 557–58.

19. Alex Keller, "The age of the projectors," *History Today* 16 (1966): 467–74; quotation on 472.

20. Keller remarks that it was England's relative technological backwardness, as compared to certain other continental European regions, that made it look as if the country was developing so quickly. Ibid., 467, 469. See also Brenda J. Buchanan, "The art and mystery of making gunpowder: The English experience in the seventeenth and eighteenth centuries," in B. D. Steele and T. Dorland, eds., *The heirs of Archimedes: Science and the art of war through the age of Enlightenment*, 235–74 (Cambridge: MIT Press, 2005), 235. Joan Day, introduction to J. Day and R. F. Tylecote, *The industrial revolution in metals*, 1–36 (London: Institute of Metals, 1991), 5–8, 22.

21. On technological project-makers at the Bureau during the period 1715–36, Lindqvist, *Technology on trial*, 135–57.

22. The main source of the following account is Bring, "Bidrag till Christopher Polhems lefnadsteckning," 3–119. Michael Lindgren has written about Polhem in *Svenskt biografiskt lexikon* and other places, but his work adds little to that of Bring. Michael Lindgren, "Polhem, Christopher," in *Svenskt biografiskt lexikon*, 29:388–93. On Laboratorium Mechanicum and Laboratorium Chymicum, Lindqvist, *Technology on trial*, 99.

23. Apart from Wrangel, he was supported by the assessors Peter Cronström and Harald Lybecker. Bring, "Bidrag till Christopher Polhems lefnadsteckning," 23. He also had the continuing support of Fabian Wrede, who was president of the Bureau of Commerce until 1712. Gerentz, *Kommerskollegium*, 63–64.

24. On Polhem, "Kongl. Maj:ts och Riksens Bergs-Collegii Relation," in [Loenbom], *Handlingar*, pt. 12; 80, 83–84, 90. Fortsättning til slut, af Kongl. Maj:ts och Riksens Bergs-Collegii Relation," in [Loenbom], *Handlingar*, pt. 13, 102–4; quotation on 105.

25. My emphasis. Karl XI to the Bureau of Mines, 6 May 1691, in Riksregistraturet, RA. As quoted in Bring, "Bidrag till Christopher Polhems lefnadsteckning," 16.

26. The word *ingenium* was also used in his leaving certificate from Uppsala, which stated that he had followed his "natural ingenium," as he had primarily studied mathematics, physics, and mechanics. Bring, "Bidrag till Christopher Polhems lefnadsteckning," 13. The word was also used to describe him in a letter of recommendation sent by the Uppsala Collegium Curiosorum to Casten Feif in Oct. 1711. Ibid., 45.

27. Ibid., 22–25, 30–36. For a complementary, although strongly hagiographic view, Lindroth, *Gruvbrytning och kopparhantering*, 331–49.

28. Patrik Winton, *Frihetstidens politiska praktik: Nätverk och offentlighet 1746–1766*, Acta Universitatis Upsaliensis Studia Historica Upsaliensia 223 (Uppsala: Uppsala Universitet, 2006), 11–12, 17, 48.

29. Polhem had been ennobled in 1716. Bring, "Bidrag till Christopher Polhems lefnadsteckning," 45–49, 57.

30. Bruce T. Moran, "Patronage and institutions: Courts, universities, and academies in Germany: An overview, 1550–1750," in B. T. Moran, ed., *Patronage and institutions: Science, technology and medicine at the European court, 1500–1750* (Woodbridge, UK: Boydell Press, 1991), 176–78.

31. Fors, "Wallerius and the laboratory of enlightenment," 7–11. H. Fors, "Matematiker mot linneaner: Konkurrerande vetenskapliga nätverk kring Torbern Bergman," in S. Widmalm, ed., *Vetenskapens sociala strukturer: Sju historiska fallstudier om konflikt, samverkan och makt*, 25–53 (Lund: Nordic Academic Press, 2008), 24–28.

32. Lindqvist, *Technology on trial*, 101.

33. On interactions between the Bureau and Uppsala, Hjalmar Fors, "Kemi, paracelsism och mekanisk filosofi: Bergskollegium och Uppsala cirka 1680–1770," *Lychnos* (2007): 211–44; and Fors, "J. G. Wallerius and the laboratory of Enlightenment," in E. Baraldi, H. Fors, and A. Houltz, eds., *Taking place: The spatial contexts of science, technology and business* (Sagamore Beach: Science History Publications, 2006), 3–33. On kinship ties at the Bureau, Lindqvist, *Technology on trial*, 217–18. On the importance of aristocratic kinship ties, Mirkka Lappalainen, "Släkt och stånd i bergskollegium före reduktionstiden," *Historisk tidskrift för Finland* 87, no. 2 (2002): 145–72. As Lappalainen remarks (169), there is no thorough study of kinship ties at the lower levels in the organization.

34. Lindqvist, *Technology on trial*, 98, 101. The Bureau of Mines was roughly twice the size of the Bureau of War; Kaiserfeld, *Krigets salt*, 58.

35. Staffan Högberg, "Inledning," in Anton von Swab, *Anton von Swabs berättelse om Avesta kronobruk 1723*, 7–30, Jernkontorets bergshistoriska skriftserie 19 (Stockholm: Jernkontoret, 1983), 12.

36. The original Latin name of Collegium Curiosorum was dropped in 1719 when the society was reestablished as the Guild of Book Learning (Bokwetts Gillet), a name that was again changed to the Uppsala Society for Science (Vetenskapssocieteten) toward the end of the 1720s.

37. Participating were, apart from the learned university librarian Benzelius Pehr Elvius and Harald Wallerius, Lars Roberg, Rudbeck the younger, and Uppmark. In addition, there were Wallerius's sons Johan and Göran. Corresponding members were Swedenborg and Polhem. Dunér, *Världsmaskinen*, 56.

38. Bring, "Bidrag till Christopher Polhems lefnadsteckning," 34, 45, 58. On Göran Wallerius's journeys and career, Lindqvist, *Technology on trial*, 175.

39. As David Dunér, puts it, "Descartes's teaching on corpuscles and theory of vortexes in Principia philosophiae was the obvious point of departure for Swedenborg and the circle around Collegium Curiosorum." Dunér, *Världsmaskinen*, 255 (my translation).

40. Ibid., 159–76, 178–79.

41. Staffan Rodhe, *Matematikens utveckling i Sverige fram till 1731* (Uppsala: Uppsala Universitet, 2002), 13–36. Dunér, *Världsmaskinen*, 79–80.

42. Bring, "Bidrag till Christopher Polhems lefnadsteckning," 62–63.

43. Ibid., 60.

44. Widmalm, "Instituting science in Sweden," 244–47.

45. Later, Mårten Triewald would have a similar role in the founding of the Swedish Royal Academy of Sciences; Lindqvist, *Technology on trial*, 183–214.

46. Swedenborg, "Document 39. Swedenborg to Ericus Benzelius," London, 13 Oct. 1710, in Tafel, *Documents concerning the life and character of Emanuel Swedenborg*, 1:206–8; quotation on 207 (Tafel's translation from Swedish original).

47. Swedenborg, "Document 40. Swedenborg to Ericus Benzelius," received 30 Apr. 1711, ibid., 1:209–12. In 1725 Swedenborg would also convince the Bureau of Mines to buy a first-grade air pump (manufactured by Francis Hauksbee). Lindqvist, *Technology on trial*, 162.

48. "Document 41. Extracts from the minutes of the Literary Society of Upsal.," in Tafel, *Documents concerning the life and character of Emanuel Swedenborg*, 1:213.

49. Swedenborg, "Document 43. Swedenborg to Ericus Benzelius," received Jan.1712, ibid., 1:219.

50. Swedenborg, "Document 46. Swedenborg to Ericus Benzelius," 8 Sept. 1714, ibid., 1:230.

51. Ibid., 1:229–33; list on 230–31; quotation on 231 (Tafel's translation).

52. Swedenborg, "Document 48. Swedenborg to Ericus Benzelius," 9 Aug. 1715, ibid., 1:236–38.

53. Polhem commented on Swedenborg's attempts to make a flying machine. "With respect to flying by artificial means, there is perhaps the same difficulty contained in it, as in making a perpetuum mobile or gold by artificial means; although, at first sight, it seems as easy to be done as it is desirable; for whatever any one approves strongly, he has generally a proportionate desire to carry it out." Polhem, "Document 61. Polhem to Swedenborg," 5 Sept. [1716], ibid., 1:269.

54. The position came with no salary. Swedenborg was made a regular assessor in 1724 (with reduced salary) and received full salary from 1730. Swedenborg, "Document 67. Swedenborg to Ericus Benzelius," [toward end of Dec., 1716], ibid., 1:273–75. "Document 155. Swedenborg's life at the college in 1724," ibid., 431–34. "Document 158. Swedenborg's life at the college from 1727 to 1732," ibid., 438–41.

55. Bring, "Bidrag till Christopher Polhems lefnadsteckning," 73. Polhem's printed literary output was diminutive in his own lifetime. Ibid., 80, 97–98. On the king's interest in *Daedalus Hyperboreus*, Swedenborg, "Document 67. Swedenborg to Ericus Benzelius," [toward end of Dec., 1716], in Tafel, *Documents concerning the life and character of Emanuel Swedenborg*, 1:274.

56. "My son the assessor, has come home, and he reports that old Hiärne cannot digest the pills he has procured for himself. Let him, in his old age, consider for once, how he has acted, and endeavor to spare himself and others, by being less abusive." Quotation from Jesper Svedberg, "Fifth letter of Bishop Swedberg to Johan Rosenadler," dated 2 Jan. 1720, in Tafel, *Documents concerning the life and character of Emanuel Swedenborg* 1:166 (Tafel's translation).

57. Swedenborg, "Document 82. Swedenborg to Benzelius," Oct. 1718, ibid., 1:303–4; quotation on 303 (Tafel's translation).

58. Swedenborg, "Document 86. Swedenborg to Benzelius," 26 Nov. 1719, ibid., 1:312–15; quotation on 313 (Tafel's translation).

59. Swedenborg, "Document 78. Swedenborg to Benzelius," 30 Jan. 1718, ibid., 1:296–98; quotation on 297 (Tafel's translation). See also Swedenborg, "Document 77. Swedenborg to Benzelius," 21 Jan. 1718, ibid., 293–95.

60. Lindqvist, *Technology on trial*, 158–63. Dunér, *Världsmaskinen*, 106, 133, 136, 150.

61. Lars Bergquist, *Swedenborgs hemlighet: Om Ordets betydelse, änglarnas liv och tjänsten hos Gud* (Stockholm: Natur och Kultur, 1999), 52, 74, 87, 90–93, 123–25. Dunér, *Världsmaskinen*, 55–61, 99–101, 125, 133, 136, 150, 240, 255. Lindqvist, *Technology on trial*, 161–63.

62. Swedenborg, "Document 101. Swedenborg to Benzelius," 26 May 1724, in Tafel, *Documents concerning the life and character of Emanuel Swedenborg*, 1:337–39. Henckel, "Document 117. Henkel, the councilor of mines, to Swedenborg," 21 Nov. 1724, ibid., 356.

63. On Henckel's influence as a teacher, Dym, *Divining science*, 84–85.

64. Swedenborg, "Document 102. Swedenborg to Benzelius," 20 Aug. 1724, in Tafel, *Documents concerning the life and character of Emanuel Swedenborg*, 1:339–40; quotation on 340 (Tafel's translation).

65. In a comment on Hiärne's controversy with Swedenborg's father, Swedenborg stated that "I read through his Chemistry, and found that he is but very little grounded in the principles upon which chemistry is founded." Swedenborg, "Document 76. Swedenborg to Benzelius," 14 Jan. 1718, ibid., 1:292. See also Swedenborg, "Document 77. Swedenborg to Benzelius," 21 Jan. 1718, ibid., 293–95.

66. Polhem was disdainful of gold making also in remarks made in essays on other topics. Bring, "Bidrag till Christopher Polhems lefnadsteckning," 98. On Swedenborg's views, see Jan Häll, *I Swedenborgs labyrint: Studier i de gustavianska swedenborgarnas liv och tänkande* (Stockholm: Atlantis, 1995), 101–5.

67. For an explanation of the underlying chemical theories, Principe *Aspiring adept*, 37–40.

68. One manuscript is a draft letter and was most likely addressed to Polhem's brother-in-law M. L. Manderström; it is dated 2 Oct. 1727. The second

manuscript is undated but was probably written some time during the 1720s. Christopher Polhem, "Mathematiskt och mechaniskt bevijs att guldmakeri artificialiter per lapidem philosophorum ähr omöjeligt I vår horizont" and "Svar om lapide philosophorum," in *Christopher Polhems efterlämnade skrifter III: Naturfilosofiska och fysikaliska skrifter redigerade av Axel Liljecrantz* (Uppsala: Almqvist och Wiksell, 1952–53), 291–98.

69. Smith, "Texture of matter as viewed by artisan, philosopher, and scientist," 14–18.

70. David Dunér, "Naturens alfabet: Polhem och Linné om växternas systematik," *Svenska linnésällskapets årsskrift* (2008): 35. Bring, "Bidrag till Christopher Polhems lefnadsteckning," 59.

71. Swedenborg, "Document 93. Swedenborg to Benzelius," 2 May 1720, in Tafel, *Documents concerning the life and character of Emanuel Swedenborg,* 1:325–26.

72. Emanuel Swedenborg, *Om järnet och de i Europa vanligaste vedertagna järnframställningssätten . . . under redaktion av Hj. Sjögren* (1734; Stockholm: Wahlström och Widstrand, 1923), xxvi–xxvi.

73. On Swedenborg's views of alchemy, see further Dunér, *Världsmaskinen,* 11, 16–17, 295–98; and Häll, *Swedenborgs labyrint,* 101–5.

74. Swedenborg, *Om järnet:* on crocus martis, 365–67; on pigment, 368–69; on medicines, 373–74, 380–82, 388–90.

75. Ibid.: on magnetism, 304; on Boyle, 403–5; on growth of iron ore, 347–48.

76. Dunér, *Världsmaskinen,* 34, 41, 88.

77. Ibid., 55–61; Rodhe, *Matematikens utveckling i Sverige,* 50–55; Sven Odén, "Brandt, Georg," in *Svenskt biografiskt lexikon,* 5:784–89.

78. For the reading of early-career works in this way, see Fors, "Matematiker mot linneaner".

79. Georg Brandt, *En grundelig anledning til mathesin universalem och algebram, efter herr And. Gabr. Duhres håldne praelectioner sammanskriven af Georg Brandt* (Stockholm: Joh. L. Horrn, 1718), xv–xvi.

80. The following is an interpretation of a passage that reads in full as follows:

> Concerning this matter, the aforementioned Johan Freind expresses himself very remarkably, claiming that those earlier writers on chymical phaenomena, have almost used such fundaments, as hardly can be understood, and can be deciphered even less [.] [H]ence it may not seem strange if they have built nothing on a foundation of nothing; because they pronounce themselves, he says, as if their intent were to hide their own ignorance, or at least as if they would not like to give others enlightenment in this matter[.] [F]or this reason all that they say about the origin and foundation of Chymistry is not only dark, but also fabrications and such as [it is, it is] neither in agreement with nature nor with their own aims[.] [T]herefore, there has supposedly been no one who has laid a foundation of chymistry, upon which one could build a correct and philosophical state-

ment about this art since they have not investigated how the mechanical nature of bodies truly should be[.] [B]ut they have rather fabricated such as one as they would have wanted it to be, and therefore have ascribed bodies the layers and properties that opposed mechanics as well as themselves. For this reason the aforementioned Joh. Freind, claims that when he wanted to teach his audience a thorough knowledge of Chymistry and its foremost operations, he has no easier and milder path to go, when doing this, than through demonstrations of geometry and physics.

I have chopped up the original passage into shorter sentences to make it more legible for contemporary English readers. In the original Swedish it is constructed as two sentences. Ibid., xvi–xvii.

81. The passage also echoes Boyle, *Sceptical chymist*, and agrees with George Castle's interpretation of it. Principe, *Aspiring adept*, 49–50. See also Robert Boyle, "The sceptical chymist," in Thomas Birch, ed., *Robert Boyle: The works* (HIldesheim: Georg Olms, 1965), 1:474–586.

82. Bergskollegium huvudarkivet, Ink. brev, suppliker, rannsakningar m.m. 1721. 2. E 4:147, letter 118, registered 22 Apr. 1721; letter 122, registered 23 June 1721, RA.

83. Bergskollegium huvudarkivet, Ink. brev, suppliker, rannsakningar m.m. 1723. 2 E 4:151, letter 418, registered 2 Nov. 1723, RA.

84. John C. Powers, *Inventing chemistry: Herman Boerhaave and the reform of the chemical arts* (Chicago: University of Chicago Press, 2012), 1–4, 192–201.

85. Jonathan Swift, foreword to *Gulliver's travels* (1726).

86. Ursula Klein, "Experimental history and Herman Boerhaave's *Chemistry of plants,*" *Studies in History and Philosophy of Science Part C: Studies in History and Philosophy of Biological and Biomedical Sciences* 34 (2003): 534–35, 537. G. A. Lindeboom, "Boerhaave, Hermann," in *Dictionary of scientific biography*, 1–2:224–28. Herman Boerhaave, *A new method of chemistry: Including the history, theory, and practice of the art*, trans. Peter Shaw, 2nd ed. (London, 1741), 155–57.

87. Powers, *Inventing chemistry*, 116.

88. Ibid., 8, 105, 116–19. For Boerhaave's definition of chemistry, ibid., 70. For his definitions of physics and mechanics, ibid., 104–5.

89. Ibid., 2, 129. Quotation from Herman Boerhaave, "Chemistry purging itself," in *Boerhaave's orations: Translated with introductions and notes*, ed. E. Kegel-Brinksgreve and A. M. Luyendijk-Elshout (Leiden: Brill, 1983), 197; as quoted in Powers, *Inventing chemistry*, 99.

90. The official title was Riksvärdie until 1725, and from 1725 Myntvärdie. Johan Axel Almquist, *Bergskollegium och Bergslagsstaterna 1637–1857: Administrativa och biografiska anteckningar*, Meddelanden från Svenska Riksarkivet 2:3 (Stockholm, 1909), 111.

91. Gösta Selling, "Myntverkets byggnader i Stockholm," in Torsten Swensson, ed., *Kungliga myntet, 1850–1950*, 31–49 (Stockholm: Nordisk Rotogravyr, 1950), 31–37.

92. Torbern Bergman, *Åminnelse-tal öfver framledne Bergs-rådet och medicinae doctoren, samt K. Vetenskaps sällskapets i Upsala, och Kongl. Academiens i Stockholm ledamot, Herr Georg Brandt* (Stockholm: Lars Salvius, 1769), 14–15.

93. Georg Brandt, Meritförteckning 1757 [X 240] UUB. Zenzén, "Studier i och rörande Bergskollegii Mineralsamling," 16–17.

94. Bergman, *Åminnelse-tal öfver . . . Georg Brandt*, 25. A comprehensive critique of alchemy can be found on 24–25.

95. D 6:3 Cronstedt, *Bref til Hr . . . om den mystiska naturkunnogheten*, 101.

96. D 1450a Föreläsningar i kemi, 135–37, 180–93; UUB.

97. Klein, "Experimental history and Herman Boerhaave's *Chemistry of plants*,"537.

98. Georg Brandt, "Experimentum quo probatur dari attractionem mercurii in aurum," *Acta literaria et scientiarum Sveciae* (1731): 1–8. G. Brandt, "De observationis arsenico," *Acta literaria et scientiarum Sveciae* (1733): 39–43.

99. Bergman, *Åminnelse-tal öfver . . . Georg Brandt*, 14–15.

100. Erik von Stockenström, *Bergsmannanäringens nytta och skötsel, förestäld uti et tal til Kongl. Svenska Vetenskapsakademien den 28 october 1749* (Stockholm, 1749), 18–22; Daniel Ekström, *Tal, om järn-förädlingens nytta och vårdande; hållit i Kongl. Vetenskapsacademien vid praesidii afläggning d. 28. apr. 1750* (Stockholm, 1750), 26; Jean Georg Lillienberg, *Tal, om svenska bergshandteringens förmåner och hinder, hållet uti Hans Kongl. Maj:ts höga närvaro, för Kongl. Vetenskaps-Academien, vid praesidii nedläggande, den 7 febr. 1776* (Stockholm, 1776), 9.

101. Fors, "Kemi, paracelsism," 211–44 ; and H. Fors, "Occult traditions and enlightened science: The Swedish Board of Mines as an intellectual environment, 1680–1760," in *Chymists and chymistry: Studies in the history of alchemy and early modern chemistry*, ed. L. Principe (Sagamore Beach: Science History Publications/USA, 2007), 239–52. Quotations from Bergman in D 1459 d Manuscripta T. Bergman, vol. 4. (Biografiska och litteraturhistoriska anteckningar om kemister m. m.); UUB; fols. 30 and 173.

102. Smith, "Interaction of science and practice in the history of metallurgy," 361.

103. Among the Swedish chemists it would be Torbern Bergman who began to nurture French contacts in earnest. See Torbern Bergman, *Torbern Bergman's foreign correspondence*, ed. Göte Carlid and Johan Nordström (Uppsala: Almqvist och Wiksell, 1965). Hedvig af Petersens, "Om Torbern Bergmans och C. W. Scheeles franska förbindelser," *Personhistorisk Tidskrift* (1928).

104. Nils Zenzén, "Förord," in Swedenborg, *Om järnet*, xvi.

105. On Réaumur's interests, see Smith, "Texture of matter as viewed by artisan, philosopher, and scientist," 21–30.

CHAPTER 5

1. Rinman, *Bergwerks lexicon*, 1:996–1002; quotation on 1002.

2. As we will see, the Bureau chemists and mineralogists were skeptical about all theories of the composition of metals. Still, such theories had made a

temporary comeback at the Bureau by the time of Rinman's writing. Rinman acknowledged the viability of two theories regarding the composition of metals: the phlogistonist position that metals were composite substances, consisting of phlogiston and a metallic calx; and a novel theory presented by Antoine Laurent Lavoisier, that metals ultimately consisted of acid, which in Rinman's interpretation united with phlogiston to form metal. Rinman, however, adhered to local tradition insofar as he rarely mentioned phlogiston, or any theories of the composition of metals. That is to say, for all practical purposes Rinman treated metals as basic species. For his discussion of Lavoisier's acid theory, see the entry on iron, *Bergwerks lexicon*, 1:875–99, 890. For his entry on phlogiston, *Bergwerks lexicon* (Stockholm: Johan A. Carlbohm, 1789), 2:254–255.

3. Russell, "Science on the fringe of Europe," 323.

4. Ian Hacking, *Historical ontology* (Cambridge: Harvard University Press, 2004), 1, 4–5, 11.

5. The term *cameralism* is rarely used in discussions of the Swedish economic thought of the period. More often, the term *mercantilism* is used. There is, however, a degree of terminological confusion here, as Swedish mercantilism had more in common with German cameralist thought than with English mercantilism, as traditionally construed. Lars Magnusson, *Merkantilism: Ett ekonomiskt tänkande formuleras* (Kristianstad: SNS förlag,1999), 243–50.

6. Lisbet Koerner, *Linnaeus: Nature and nation* (Cambridge: Harvard University Press, 1999), 3. Koerner has called the cameralist model of society a "concept of local modernity," by which she claims that cameralists upheld an ideal that states should be autarkies, self-contained entities with no need of foreign trade, and with no commercial bonds except to their own subjects. The apparent isolationist economic policies of cameralists, however, did not mean that they failed to see the value of other types of exchanges with other states, and although some cameralists were adherents of ideals such as those described by Koerner, many were not. In a sense, my ambition to use the concept of circulation to show how the European north was an integrated part of pan-European exchanges runs contrary to that of Koerner, who paints a picture of Sweden and the Baltic region as a backwater whose inhabitants drew on the cameralist corpus to nurture their imagined splendid isolation. Ibid., 16, 34–35, 95–97, 163, 190, 193; quotation on 1.

7. Bartels, "Production of silver, copper, and lead," 78–79.Bartels, *Vom früneuzeitliche Montangewerbe zur Bergbauindustrie*, 446–47. Weber, *Innovationen im frühindustriellen deutschen Bergbau und Hüttenwesen*, 107.

8. Konečný, "Hybrid expert in the 'Bergstaat,'" 346.

9. Wakefield, *Disordered police state*, 20.

10. Dirk Hoerder, *Cultures in contact: World migrations in the second millennium* (Durham, NC: Duke University Press, 2002), 83, 85–86. Christoph Bartels, "Der Bergbau vor der hochindustriellen Zeit—Ein Überblick," in R. Slotta and C. Bartels, eds., *Meisterwerke bergbaulicher Kunst vom 13. bis 19. Jahrhundert* (Bochum: Deutchen Bergbau-Museum, 1990). Bartels, *Vom früneuzeitliche Montangewerbe zur Bergbauindustrie*, 274. Bartels "Production of silver, copper, and lead," 80, 86. Bjørn Ivar Berg, *Gruveteknikk ved*

Kongsberg Sølvverk 1623–1914 (Dragvoll: Norges teknisk-naturvitenskapelige universitet, 1998); Donata Brianta, "Education and training in the mining industry, 1750–1860: European models and the Italian case," *Annals of Science* 57 (2000): 268–75. Weber, *Innovationen im frühindustriellen deutschen Bergbau und Hüttenwesen*, 47. For an enlightening discussion on the global spread and use of assaying equipment manufactured in central Europe, i.e., Hessian and Bavarian crucibles, see Marcos Martinón-Torres, "The tools of the chymist: Archaeological and scientific analyses of early modern laboratories," in Principe, *Chymists and chymistry*, 149–63.

11. Bartels, "Production of silver, copper, and lead," 81. Dym, *Divining science*, 79–80.

12. On the importance of German artisans and administrators for the Swedish mining administration, see Almquist, *Bergskollegium*, 9–15. Lindroth, *Gruvbrytning och kopparhantering*, 43, 63–66, 153, 279–81. Lappalainen, "Släkt och stånd," 145–72, esp. 154, 157–58.

13. Eli Heckscher, *Sveriges ekonomiska historia från Gustav Vasa: Före frihetstiden*, pt. 1:2 (Stockholm: Bonniers, 1936), 465–70. Hildebrand, *Swedish iron*, 98–100.

14. On Falun, see Zenzén, "Från den tid, då vi skulle transmutera järn till koppar," 88–151. Lindqvist, *Technology on trial*, 135–57. On Sala (a silver mine), see Petrus Norberg, *Sala gruvas historia under 1500- och 1600-talen* (Sala: Sala-postens boktryckeri AB, 1978), 431.

15. Rinman, *Bergwerks lexicon*,2:1249–72.

16. Almquist, *Bergskollegium*, 9–15. Lindroth, *Gruvbrytning och kopparhantering*, 153–57. Martin Fritz, "The economic background," in Fritz, *Iron and steel on the European market*, 19. Lappalainen, "Släkt och stånd," 154, 157–58. Lappalainen, *Släkten—makten—staten*, 44–47.

17. Bengt Hildebrand, *Kungl. Svenska vetenskaps akademien: Förhistoria, grundläggning och första organisation* (Stockholm: K. Vetenskapsakademien, 1939), 616.

18. Johann Heinrich Gottlob von Justi, *Grundriss des gesamten Mineralreiches; worinnen alle Fossilien in einem, ihren wesentlichen Beschaffenheiten gemässen, Zusammenhange vorgestellet und beschrieben werden* (Göttingen, 1757), [iii–iv], 75–76, 102–5; quotation from unpag. dedication.

19. Mathias Persson, *Det nära främmande: Svensk lärdom och politik i en tysk tidning, 1753–1792* (Uppsala: Uppsala Universitet, 2009), 100, 105–13.

20. Hansen, as quoted in Weber, *Innovationen im frühindustriellen deutschen Bergbau und Hüttenwesen*. 54n33. Original in OBA Clausthal Fach 767 Nr. 87: Bericht Hansens über die Ergebnisse seiner Schwedenreise, 3.

21. Interesting details of exchanges between, in particular, Uppsala and the Mining Academy in Schemnitz from a somewhat later period can be found in Konečný, "Hybrid expert," 337–39, 342.

22. Weber, *Innovationen im frühindustriellen deutschen Bergbau und Hüttenwesen*, 47–49.

23. Bartels, *Vom früneuzeitliche Montangewerbe zur Bergbauindustrie*, 93, 318–22, 372. See also Lindroth, *Gruvbrytning och kopparhantering*, 340–41. Lindroth, *Christopher Polhem och Stora Kopparberget: Ett bidrag till bergs-*

mekanikens historia (Uppsala: Almqvist och Wiksell, 1951), 92–101. For a more comprehensive discussion of the industrialization of the Oberharz, see Bartels, *Vom frühneuzeitliche Montangewerbe zur Bergbauindustrie*, 373–75, 473–76.

24. Johann Heinrich Pott, *Fortsetzung dererer Chymischen Untersuchungen, welche von der Lithogeognosie oder Erkäntniss und Bearbeitung derer Steine und Erden specieller handeln* (Berlin: Christian Friedrich Voss, 1751), 2–5.

25. His mineralogy was originally published anonymously. [Axel Fredrik Cronstedt], *Försök til Mineralogie, eller mineral-rikets upställning* (Stockholm, 1758), iv.

26. Johann Heinrich Pott, *Send-Schreiben an den Herrn Berg-Rath von Justi darin die Einwürfe, die er ihm in seinen wieder aufgelegten Chymischen Schriften von neuen gemacht hat, erörtert und abgelehnet, und die darinn angefochtene Chymisch-Physicalische Materien weiter untersucht und ausgeführt werden* (Berlin: Spenerischen Buchhandlung, 1760), 16–17.

27. Axel Fredrik Cronstedt, *Cronstedts Versuch einer Mineralogie: Vermehret durch Brünnich* (Copenhagen: Prost und Rothens Erben, 1770). Axel Fredrik Cronstedt, *Axel von Kronstedts Versuch einer Mineralogie: Aus neue as dem Schwedischen übersetzt und nächst verschiedenen Anmerkungen vorzüglich mit äussern Beschreibungen der Fossilien vermehrt von Abraham Gottlob Werner* (Leipzig: Siegfried Lebrecht Crusius, 1780), vi. One can also mention F. W. H. von Trebra's criticism of the same work. Dym, *Divining science*, 179–80. See also Eddy, who has noted the influence of Wallerius and Cronstedt on Werner's writings. Eddy, *Language of mineralogy*, 127.

28. The recent work of Konečný is an exception. Konečný, "Hybrid expert," 337–39, 342, 346.

29. Weber, *Innovationen im frühindustriellen deutschen Bergbau und Hüttenwesen*. On Heynitz's journey to Sweden, 47–57; on his suggestions for reforms and inspirations, 87, 131, 133, 237–38; for Weber's conclusions, 242.

30. Baumgärtel, *Bergbau und Absolutismus*, 54.

31. Lindroth, *Gruvbrytning och kopparhantering*, 331. Lindroth then contradicts himself three pages later, on 334–35. He did indeed acknowledge a strong German influence on Swedish mining. It is the nationalistic tendency in his writings that I wish to highlight, as when he forgets to acknowledge the late seventeenth-century foreign smelters and chymists engaged in Falun—including Johann Kunckel—as experts. Presumably their status as both foreigners and "alchemists" made them default charlatans. Ibid., 280.

32. Axel Fredrik Cronstedt, *Inträdes-tal om medel till mineralogiens vidare förkofran: Hållit för Kongl. Svenska Vetenskapsakademien den 9 Feb. 1754* (Stockholm: Lars Salvius, 1754), 13.

33. Allen G. Debus, "Fire analysis and the elements in the sixteenth and seventeenth centuries," *Annals of Science* 23 (1967): 133–39.

34. Hiärne, *En ganska liten bergslykta*, 28–29. A similar view was also held by Bromell. Magnus von Bromell, *Mineralogia: Eller inledning til nödig kundskap at igenkiänna och upfinna allahanda berg-arter, mineralier, metaller samt fossilier*, 2nd ed. (Stockholm: Gottfried Kiesewetter, 1739), 54.

35. Powers, *Inventing chemistry*, 37, 67, 74, 149, 171–74. Boerhaave became more critical of metallic transmutation in the 1730s (*Elementa chemiae*

had been composed in 1728–29). The more critical view was published in the *Transactions of the Royal Society* in 1733–34, 1736–37. Ibid., 180–91.

36. Boyle, "Sceptical chymist," 475–76.

37. Frank Greenaway, "Pott, Johann Heinrich," in C. C. Gillispie, ed., *Dictionary of scientific biography*, 11:109.

38. Pott, *Fortsetzung dererer Chymischen Untersuchungen*, iii.

39. Georg Brandt, "Dissertatio de semi-metallis," *Acta literaria et scientiarum Sveciae* (1735): 1–12; quotation on 1. Presumably following Brandt, similar definitions could be found in the works of Cronstedt and Bergman. Axel Fredrik Cronstedt, *An essay towards a system of mineralogy: Translated from the original Swedish, with notes by Gustav von Engestrom. To which is added, a treatise on the pocket-laboratory, containing an easy method, used by the author, for trying mineral bodies, written by the translator. The whole revised and corrected, with some additional notes by Emanuel Mendes Da Costa* (London, 1770), 235. Torbern Bergman, *Anledning til föreläsningar öfver chemiens beskaffenhet och nytta, samt naturliga kroppars almännaste skiljaktigheter* (Stockholm: Uppsala och Åbo, 1779), 45.

40. "Ex definitione semi-metalli supra-memorata patet, vitriola, cinnabarim, minerasque vel venas metalliferas, item terras & vitra hujus generis, in quibus nihil puri (*gediget*) [Swedish: solid or pure (metal)] inest, pro semi-metallis, aut metallis merito haberi non posse." Brandt, "Dissertatio de semi-metallis," 1–12; quotation on 2. I thank Maria Berggren at Uppsala University library for her help with this translation.

41. X277 Prober-konsten sammanskrefwen af Jacob Fischer proberare; KB. Manuscript course overview written by Lars Schultze.

42. In a recent paper Newman places the origins of this definition in medieval alchemy and ultimately in the corpus of Aristotle. He argues that it was transmitted into eighteenth-century chemistry through Robert Boyle, whose usage in turn derives directly from Daniel Sennert. Newman remarks rightly that Antoine Laurent Lavoisier was not the author of the negative-empirical concept (on 262–63). Yet Newman seems unaware of his paper's important contribution to a long-standing discussion among historians of eighteenth-century chemistry. Other historians, such as Cassebaum and Kauffman, have argued that Carl Wilhelm Scheele and Torbern Bergman were the authors of the concept. Porter has argued that Scheele and Bergman proceeded from earlier usages at the Swedish Bureau of Mines, and discusses Cronstedt, before he finally decides that the German chemist Johann Heinrich Pott was the likely author. William R. Newman, "The significance of 'chymical atomism," in Edith Dudley Sylla and William R. Newman, eds., *Evidence and interpretation in studies on early science and medicine*, 248–64 (Leiden: Brill, 2009), 248–57. Cassebaum and Kauffman "Analytical concept of a chemical element in the work of Bergman and Scheele," 447–56. Porter, "Promotion of mining and the advancement of science," 549–50, 557–58. See also Fors, "Kemi, paracelsism och mekanisk filosofi," 180, as well as Robert Siegfried and Betty Jo Dobbs, "Composition: A neglected aspect of the chemical revolution," *Annals of Science* 24 (1968): 281–84, 292–93.

43. Newman, "Significance of 'chymical atomism,'" 254.

44. Brandt, "Dissertatio de semi-metallis," 4–6.

45. Ibid. Axel Fredrik Cronstedt, "Rön och försök gjorde med en malm-art, från Los kobolt grufvor i Färila Socken och Helsingeland," *Kungliga Vetenskapsakademiens Handlingar* (1751). A. Cronstedt, "Fortsättning af rön och försök, gjorde med en malm-art från Los kobolt-grufvor," *Kungliga Vetenskapsakademiens Handlingar* (1754).

46. The metal was known to the indigenous population of South America before the arrival of Europeans, and after that it was known by knowledgeable assayers in the Spanish colonies. Eventually it came to the attention of European men of science, and discussions of whether it was a new noble metal had begun in the 1740s. Henric Theophil Scheffer, "Det hvita gullet, eller sjunde metallen. Kalladt i Spanien Platina del Pinto, Pintos små silfver, beskrifvit til sin natur," *Kungliga Vetenskapsakademiens Handlingar* (1752): 269–75; H. Scheffer, "Tilläggning om samma metall," *Kungliga Vetenskapsakademiens Handlingar* (1752): 276–78; L. B. Hunt, "Swedish contributions to the discovery of platinum: The researches of Scheffer and Bergman," *Platinum Metals Review* 24 (1980): 31–36; A. Galán and R. Moreno, "Platinum in the eighteenth century: A further Spanish contribution to an understanding of its discovery and early metallurgy," *Platinum Metals Review* 36 (1992): 40–47; Luis Fermin Capitan Vallvey, "The Spanish monopoly of Platina: Stages in the development and implementation of a policy," *Platinum Metals Review* 38 (1994): 22–25.

47. Bengt Quist, "Rön om bly-erts," *Kungliga Vetenskapsakademiens Handlingar* (1754). Sten Lindroth, *Kungliga Svenska vetenskapsakademiens historia 1739–1818*, vol. 1, pt. 1 (Stockholm: Kungl. Vetenskapsakademien, 1967), 526.

48. Cronstedt, "Rön och försök," 291–92.

49. Ibid., 291; Cronstedt, "Fortsättning af rön och försök," 44–45; Cronstedt, "Några rön och anmärkningar vid Platina di Pinto," *Kungliga Vetenskapsakademiens Handlingar* (1764): 222–26. Scheffer, "Det hvita gullet," 273. Georg Brandt, *Tal om färg-cobolter, hållit för Kongl. Vet. Acad. Vid praesidii nedläggande den 30 jul. 1760* (Stockholm: Lars Salvius, 1760), 8–9, 13, 19.

50. D 1450a Föreläsningar i kemi, 180–85; quotation on 183; UUB. Bergman, too, strongly emphasized the importance of analysis and synthesis. Torbern Bergman, *Physical and chemical essays*, trans. from the original Latin by Edmund Cullen, to which are added notes and illustrations, by the translator (J. Murray: London, 1784), 1:xxv.

51. D 1450a Föreläsningar i kemi, 185–93; quotations on 187; UUB. "Proven causes" is an interpretation of the original "demonstrative skiähl" (on 187).

52. Axel Fredrik Cronstedts arkiv [F1:2] (Vetenskapliga anteckningar och manuskript, vol. 2) "Docimastica et chemica af Brandt," 431–60; on 442–43, KVA.

53. Justi, *Grundriss des gesamten Mineralreiches*, 75–76, 102–5; quotation on 105.

54. Lindroth, *Kungliga Svenska vetenskapsakademiens historia*, vol. 1, pt. 1, 519.

55. Cronstedt, *Essay*, xi.

56. Patriotic feelings and the coordinated efforts to produce a response are clear from, e.g., Pehr Wilhelm Wargenting to Axel Fredrik Cronstedt, 29 May 1760, 31 July 1760. In Axel Fredrik Cronstedts arkiv [E1:1] (Manuskript, vol. 8), KVA.

57. Persson, *Det nära främmande*, 100.

58. For an overview of mineralogical systems of the eighteenth century, Rachel Laudan, *From mineralogy to geology: The foundations of a science, 1650–1830* (Chicago: University of Chicago Press, 1987), 23–25.

59. This was one of the reasons for his criticism of Woltersdorf (see above), and Pott, *Fortsetzung dererer Chymischen Untersuchungen*, 3–4.

60. There were also personal conflicts. Especially Wallerius and Bergman disliked the competition from Linnaeus. Wallerius to Cronstedt, 17 Mar. 1760, [MS Cronstedt, vol. 8], KVA, Bergman to Bengt Bergius, 21 Apr.1769, Bergianska Brevsaml. 13, 778.

61. Johan Gottschalk Wallerius, *Mineralogia eller mineralriket indelt och beskrifwit* (Stockholm, 1747). Porter "Promotion of mining and the advancement of science," 553. For a survey of the Swedish tradition of mineralogical classification in the eighteenth century, see Lundgren, "Bergshantering och kemi i Sverige under 1700-talet," *Med Hammare och Fackla* 29 (1985): 103–20. William A. Cole, *Chemical literature, 1700–1860: A bibliography with annotations, detailed descriptions, comparisons and locations* (London, 1990), 554–55.

62. To Wallerius, chemistry was a systematic science, and the theoretical foundation of a number of practical applications such as assaying, pharmacy, metallurgy, dyeing, and so on. Wallerius, "Bref om chemiens rätta beskaffenhet nytta och wärde" (Stockholm, 1751); Fors, *Mutual favours*, chap. 3; Fors, "Laboratory of Enlightenment," 3–14, 19–21; H. Fors, "Vetenskap i alkemins gränsland: Om J. G. Wallerius *Wattu-riket*," *Svenska Linnésällskapets Årsskrift* (1996–97): 54–57.

63. [Cronstedt], *Försök til mineralogie*. Sven Rinman, *Åminnelsetal öfver framledne Bergmästaren ... Cronstedt* (Stockholm: Lars Salvius, 1766), 10–11. Lundgren, "Bergshantering och kemi," 105–14.

64. Cronstedt, *Essay*, xii.

65. Ibid., xiii.

66. Cronstedt, however, also hazarded a guess, "because we have strong reasons to believe that the calcareous and argillaceous earths are the two principal ones, of which all the rest are compounded, although this cannot yet be perfectly proved to a demonstration." Ibid., xvii–xviii.

67. Ibid., 241. The quotation has been lightly edited for spelling. The use of the analytical concept for the purposes of discovery also comes across in his writings on nickel, as in the following passage: "I have not, besides the nickel, found any metal or metallic composition, which 1. Becomes green when calcined. 2. Yields a vitriol, whose colcothar also becomes green in the fire. 3. So easily unites with sulfur, and forms with it a regule of such a peculiar nature, as the nickel does in this circumstance; and that 4. Does not unite with silver, but only adheres or sticks close to it, when they have been melted together." Ibid., 240.

CHAPTER 6

1. My crude English interpretation does not do justice to the original Swedish verse. It reads: "Om det är sant, hwad Locken sagt, / At Gud materien kunnat gifwa / Förmögenhet til tankemagt, / Och såleds inga andar blifwa: / Så är det ju hans tanke-slut, / At hwad wi själ och anda kalla / Med döden lärer sönderfalla, / Och med wår Lifstid löpa ut. " Hedvig Charlotta Nordenflycht, "Wigtiga frågor til en Lärd, med Auctorens egit svar," in H. Nordenflycht, *Qvinligit tanke-spel: af en herdinna i Norden* (1744), 47. As quoted by Martin Lamm, *Upplysningstidens romantik: Den mystiskt sentimentala strömningen i svensk litteratur*, pt. 1 (1918; Lund, 1963), 169.

2. Lamm, *Upplysningstidens romantik*, 152–53.

3. Dunér, *Världsmaskinen*, 205.

4. Conway, *Principles of the most ancient and modern philosophy*, 20, 29. For Conway's criticism of Descartes, ibid., 148–49.

5. Ibid., 117–18.

6. Dunér, *Världsmaskinen*, 205.

7. Copenhaver, "Natural magic, hermetism, and occultism in early modern science," 269.

8. D 6:3 Cronstedt, *Bref til Hr . . . om den mystiska naturkunnogheten*, 70–71. RA.

9. On Jews, see ibid., 68, 80.

10. Ibid., 74.

11. Ibid., 75–78, 83–84.

12. Ibid., 80–81, 86–87.

13. Ibid., 91, 99.

14. Ibid., 91.

15. Dym, *Divining science*, 3, 6, 13, 116–17, 119. Dym emphasizes the Freiberg dowsers' status as trusted locals. A trustworthy dowser was a local man with good knowledge of mining. He made modest claims, was available, forthcoming for examination, submissive to authority, and conducted himself with evident humility. If he exhibited these traits he was accepted in the mining community as a useful specialist of middling status and could earn a wage on a par with that of a foreman. On the other hand, individuals regarded as "crafty, elusive, or on the move" were not to be trusted. Ibid., 134–35.

16. D 6:3 Cronstedt, *Bref til Hr . . . om den mystiska naturkunnogheten*, 98–99. RA. This dowser's behavior (as narrated) would have made him seem a suspicious individual in Freiberg, the capital of dowsing. Dym, *Divining science*, 105, 124–32, 134–35.

17. Annie Mattsson, *Komediant och riksförrädare: Handskriftcirkulerade smädeskrifter mot Gustaf III* (Uppsala: Uppsala Universitet, 2010), 111–12.

18. There is evidence of this attitude among several Swedish eighteenth-century mining officials and chemists. On Hiärne's complaints about his German assistants and the many German apothecaries in Stockholm, see Lindroth, "Hiärne, Urban," 85–86. See also letter from Sven Rinman to Torbern Bergman, 26 Jan. 1769, in G21 UUB, and letter from Bergman to Johan Gottlieb Gahn, 3 Aug. 1772, in Jan Trofast, ed., *Johan Gottlieb Gahn Brev: Utgivna med kommentarer av Jan Trofast*, pt. 2 (Lund: Jan Trofast, 1994). For some functions

of this type of remark in eighteenth-century enlightened scientific discourse, see Fors, *Mutual favours*, 99–105.

19. Samuel Troili to Bergskollegium, 5 Feb. 1747 (registered 9 Feb. 1747) in Riksdagen, 1746–47, Bergsdeputationen, vol. 2 [R 2888], fol. 1592 b, RA. Fifty years later, Berndtson used the same episode to cast doubt on the usefulness of foreign artisans. Bernhard Berndtson, *Tal, om en bergsmans kunskaper och gjöromål hållet för Kongl. Vetenskaps Academien vid praesidii nedläggande den 1 februar. 1797* (Stockholm: Joh. P. Lindh, 1797), 35.

20. Daniel Tilas, "Till Herr Axel Fredrich Cronstedt tilläggning til des historie om mystiska naturkunnigheten," Stockholm, June 1758, in Bergskollegiums arkiv, huvudarkivet D 6:3, RA, 114–17.

21. D 6:3 Cronstedt, *Bref til Hr . . . om den mystiska naturkunnogheten*, 95–97. RA.

22. Häll, *Skogsrået, näcken och djävulen*, 320, 364.

23. Tillhagen, *Järnet och människorna*, 104–5, 107, 111, 113, 124.

24. Olsson, *Falun mine*, 6–8. On the lady of the Sala silver mine, Norberg, *Sala gruvas historia*, 394–96.

25. Dym, *Divining science*, 91.

26. Samuel Gustaf Hermelin to Torbern Bergman, 3 Mar. 1768, in G21 UUB.

27. Dym, *Divining science*, 132, 164–66.

28. Tilas, "Till Herr Axel Fredrich Cronstedt tilläggning til des historie om mystiska naturkunnigheten," Stockholm, June 1758, in Bergskollegiums arkiv, huvudarkivet D 6:3, RA,118–20.

29. Daniel Tilas, "Curriculum vitae 1–2, 1712–1757 samt fragment av dagbok september—oktober 1767," in *Historiska handlingar*, vol. 38, pt. 1 (Stockholm: Norstedts, 1966), 147, 148–49, 169–70, 200–201, 224–26.

30. For a parallel case, eighteenth-century gendering of mermaids, see Chaiklin, "Simian amphibians," 241–73.

31. D 6:3 Cronstedt, *Bref til Hr . . . om den mystiska naturkunnogheten*, 103. RA.

32. Ibid., 101.

33. On Bonde, see G. Carlquist, "Gustav Bonde," in *Svenskt biografiskt lexikon*, 5:362–77, and Carl Michael Edenborg, *Gull och mull* (Lund: Ellerströms, 1997). A discussion of Daniel Tilas, *Åminnelse-tal öfver . . . Gustav Bonde . . . Hållit i Stora Riddarehus-Salen Den 7. Junii 1766* (Stockholm: Lars Salvius, 1766), 1–47, and of Bonde is found in Hjalmar Fors, "Speaking about the other ones: Swedish chemists on alchemy, c. 1730–70"; J. R. Bertomeu-Sánchez, D. T. Burns, and B. Van Tiggelen, eds., *Neighbours and territories: The evolving identity of chemistry proceedings of the 6th International Conference on the History of Chemistry* (Leuven, 2008), 283–89.

34. Fr. Wilh. Ehrenheim, ed., *Tessin och Tessiniana* (Stockholm: Johan Imnelius, 1819), 382.

35. Lindqvist, *Technology on trial*, 136–37.

36. I.e., Johan Arckenholtz, Ulrik Rudenschöld, Fredrik Lorenz Bonde, and possibly also Carl Fredrik Nordenskiöld. Of these, Oelreich and Rudenschöld were closest to Bonde, and he kept up a regular correspondence with Arckenholtz. Carl Michael Edenborg, *Gull och mull* (Lund: Ellerströms, 1997), 160–

81; Edenborg states that Ulrik Rudenschöld had been an assessor at the Bureau, but this seems not to have been the case. Ibid., 162.

37. Lindqvist, *Technology on trial*, 136–37.

38. Gustav Bonde to Fredrik Lorenz Bonde, 1752, in Carl Trolle-Bonde, *Riksrådet grefve Gustav Bonde 2: Gustaf Bondes Litterära verksamhet*, Anteckningar om Bondesläkten (Lund: Berlingska, 1897), 150–52; quotation on 151.

39. Edenborg, *Gull och mull*. Almquist, *Bergskollegium*, 174.

40. Undated, unsigned letter [from J. G. Wallerius, probably written in 1769] to Torbern Bergman, in *Svensk brevväxling till Torbern Bergman*, G21 UUB, 675.

41. Powers, *Inventing chemistry*, 180–81.

42. Sten Lindroth, *Svensk Lärdomshistoria: Frihetstiden* (Stockholm: Norstedt, 1975), 402–3. Tore Frängsmyr , *Sökandet efter upplysningen: En essä om 1700-talets svenska kulturdebatt* (Höganäs: Wiken, 1993), 105. For a more exhaustive critique of the positions of these two scholars, see Fors, "Kemi, paracelsism," n. 72.

43. See Fors, "Vetenskap i alkemins gränsland," (1996–97): 33–60. Meinel, "Theory or practice?" 126–29; J. R. Partington, *A history of chemistry* (London, 1962), 3:169–72; Guerlac, "Some French antecedents of the chemical revolution," 100–101; Hugo Olsson, *Kemiens historia i Sverige intill år 1800* (Uppsala, 1971), 108–15, esp. 113–14. Nils Zenzén, "Johan Gottschalk Wallerius, 1709–1785, and Axel Fredrik Cronstedt, 1722–1765," in S. Lindroth, ed., *Swedish men of science, 1650–1950*, 92–97 (Stockholm: Almqvist och Wiksell, 1952).

44. Bergquist, *Swedenborgs hemlighet*, 179.

45. The phenomenon is today considered to be a rising of the land following the receding of the ice which covered Scandinavia during the last ice age.

46. Tore Frängsmyr, "The Enlightenment in Sweden," in Porter and Teich, *Enlightenment in national context*, 171.

47. Åkerman, *Fenixelden*, 266, 269, 273. For an overview of Bonde's esoteric library, ibid., 260–66. See also Bergquist, *Swedenborgs hemlighet*, 415. Bonde to Swedenborg, 7 Aug. 1760, Swedenborg to Bonde, 11 Aug. 1760, in Tafel, *Documents concerning the life and character of Emanuel Swedenborg*, 3:230–33.

48. Carl Trolle-Bonde, *Riksrådet grefve Gustav Bonde 2*, 95.

49. Ibid., 97.

50. Harry Lenhammar, *Tolerans och bekännelsetvång: Studier i den svenska swedenborgianismen 1765–95* (Uppsala, 1966), 21, 26. Bergquist, *Swedenborgs hemlighet*, 415.

51. As indicated by the previous passages, debates were mostly conducted in a private setting. Lindroth, in a 1957 article, identifies no more than a few Swedish public debates touching on these issues: in particular Hiärne's debate with M. G. von Block on prophesies and Paracelsianism, and a debate concerning the possibility of transmuting oats into rye, which began in 1757 and continued into the 1760s. Sten Lindroth, "Naturvetenskaperna och kulturkampen under frihetstiden," *Lychnos* (1957): 181–93; 185–87.

52. Tilas to Cronstedt, 16 Mar. 1760 in, Tafel, *Documents concerning the life and character of Emanuel Swedenborg*, 3:395–96 (Tafel's translation).

53. Ibid.

54. Signe Toksvig, *Emanuel Swedenborg: Scientist and mystic* (New Haven: Yale University Press, 1948), 7–12, 201. By 1763 at the latest, it was old news for everyone. Bergquist notes two printed references from 1763 that link Swedenborg to his religious works in passing. Bergquist, *Swedenborgs hemlighet*, 415.

55. Tilas to Cronstedt, 16 Mar. 1760, in Tafel, *Documents concerning the life and character of Emanuel Swedenborg*, 3:395–96 (Tafel's translation).

56. Winton, *Frihetstidens politiska praktik*, 41–42.

57. Axel Fredrik Cronstedt, *Åminnelsetal Öfver . . . Henric Teophil Scheffer . . . Hållit i Stora Riddarehus-Salen, Den 17. September 1760* (Stockholm: Lars Salvius, 1760), 1–31: 6.

58. Cronstedt, *Åminnelsetal . . . Scheffer*, 13.

59. Ibid., 16. Compare the more refined position he defended privately two years earlier. D 6:3 Cronstedt, *Bref til Hr . . . om den mystiska naturkunnogheten*, 99–100. RA.

60. Cronstedt, *Åminnelsetal . . . Scheffer*, 14.

61. Ibid., 18.

62. Bergman, *Åminnelse-tal öfver . . . Georg Brandt*, 25.

63. Bergman, *Åminnelsetal öfver . . . Anton von Swab. . . . Hållit i Stora Riddarhus-Salen Den 29 Junii 1768* (Stockholm: Lars Salvius, 1768), 1–54; quotation on 27. Compare his much more cautious position of two years earlier. Torbern Bergman, *Physisk beskrifning öfver jord-klotet: på cosmographiska sällskapets vägnar författad* (Uppsala, 1766), 121.

64. "Wrede, Henrik Jakob," 752, in Herman Hofberg, *Svenskt biografiskt handlexikon: Alfabetiskt ordnade lefnadsteckningar af Sveriges namnkunniga män och kvinnor från reformationen till nuvarande tid* 2 (Stockholm: Bonniers,1906).

65. Hjalmar Fors, "Stockenström, Erik von," in *Svenskt biografiskt lexikon*, 164:548–53; Stille-Strandell, ed. Åsa Karlsson (Stockholm, 2010), 548–53.

66. The 1780s saw a second flair of public debate, centered around the satirist Kellgren. Now, the scientific practitioners were quiet. It does seem that the issue of chemistry's relationship to alchemy was already closed. Gustav von Engeström, director of the Bureau's Laboratory, 1768–94, did not seem to have any particular emotions attached to alchemy. As is apparent from an oration on "difficulties in chemistry" held at the Swedish Royal Academy of Sciences in 1782, his knowledge of alchemy seems to be gleaned purely from books, and he discussed it as a literary tradition. Gustav von Engeström, *Tal, om vissa svårigheter och andra omständigheter, som möta vid utöfvandet af chymien; Hållet för Kongl. Vetensk. Academien* (Stockholm, 1782), 12–16. Bergman, too, previously a staunch critic, seems to have mellowed, and developed a more positive attitude to the alchemical trials found in the older chymical literature. This is evident from Bergman's D 1459 d Manuscripta T. Bergman, vol. 4. (Biografiska och litteraturhistoriska anteckningar om kemister m. m.); UUB.

67. Tilas, *Åminnelse-tal öfver . . . Gustav Bonde*, 30.

68. Ibid., 30.

69. Ibid., 30–31.

70. Ibid., 31. The Clavicula was printed in Marburg in 1747. For a discussion of this work, Åkerman, *Fenixelden*, 256–60.

71. Edenborg even includes the word *monstrous* in the full (baroque-length) title of his biography of Bonde. See Edenborg, *Gull och mull.*

72. For a recent but highly traditional treatment of Linnaeus's allegedly rural superstitions and antiquated natural philosophy, see Koerner, *Linnaeus*, 16, 93–94, 110. For a productive reinterpretation of his views of fate and divine retribution, Savin, *Fortunas klädnader*, 155–56.

73. Dunér, *Världsmaskinen*, 36. See also Lindroth, "Linné—Legend och verklighet," *Lychnos* (1965–66): 56–122; 69, 106, 108. Lars Bergquist, "Kusin-fejden: Swedenborg kontra Linné," in E. Pierre and A. W. Johansson, eds., *Över vida fält: Studier i utrikespolitik, diplomati och historia Vänbok till Mats Bergquist* (Lund: Sekel, 2008), 37–45.

74. Furthermore, Svedberg himself carried considerable influence in Swedish religious life. Bergquist, *Swedenborgs hemlighet*, 23, 30–31; on Luther, ibid., 376.

75. Lamm, *Upplysningstidens romantik*, 99. On angels' averting of disasters, Savin, *Fortunas klädnader*, 99–103, 114–26.

76. In comparison, there are a surprising number of similarities between the anecdotes that circulated around Tollstadius and Swedenborg.

77. Immanuel Kant, *En andeskådares drömmar: I ljuset av metafysikens drömmar* översättning med inledning Efraim Briem (Lund: Gleerup, 1921). On Swedenborg as a witness, 51.

78. Shapin, *Social history of truth*, 20–22, 35–38; quotation on 36.

79. Michael Heinrichs, *Emanuel Swedenborg in Deutschland: Eine kritische Darstellung der Rezeption des schwedischen Visionärs im 18. und 19. Jahrhundert*, Europäische Hochschulschriften: Reihe 20, Philosophie; Band 47 (Frankfurt am Main: Peter D. Lang, 1979), 68, 115. Gottlieb Florschütz, *Swedenborgs verborgene Wirkung auf Kant: Swedenborg und die okkulten Phänomene aus der Sicht von Kant und Schopenhauer*, Epistemata, Würzburger Wissenschaftliche Schriften, Reihe Philosophie, Band 106 (Würzburg: Königshausen und Neumann, 1992), 21. Friedemann Stengel, *Aufklärung bis zum Himmel: Emanuel Swedenborg im Kontext der Theologie und Philosophie des 18 Jahrhunderts*, Habilitationsschrift der Theologischen Fakultät der Ruprecht-Karls-Universität (Heidelberg, 2009).

80. Franco Venturi, *Italy and the Enlightenment: Studies in a Cosmopolitan Century*, ed. and with an introduction by Stuart Woolf (Longman: London, 1972), 103, 122–33.

81. For a similar conclusion, Savin, *Fortunas klädnader*, 15, 17, 23.

CHAPTER 7

1. Johann Heinrich Jung-Stilling, *Theorie der Geister-kunde* (1808); Swedish trans.: *Andelära framställd uti et med naturen, förnuftet och uppenbarelsen*

enligt swar på den frågan: hvad bör man tro och icke tro om aningar, syner och andars uppenbarelser (Gothenburg, 1812), 6–10, 19–34. Jung had also been a teacher of cameralism at Lautern (present -day Kaiserslauten). Wakefield, *Disordered police state*, 115–23.

2. Compare Israel, *Radical enlightenment*, and Israel, *Enlightenment contested*.

3. Klein and Spary, introduction to *Materials and expertise in early modern Europe*, 1, 4–5.

4. Copenhaver, "Natural magic, hermetism, and occultism in early modern science," 269–71; quotation on 290.

5. For an overview, see, e.g. Dorinda Outram, *The Enlightenment* (Cambridge: Cambridge University Press, 2005), 24–26.

6. The disenchantment of nature is actually an eighteenth-century expression made popular by Max Weber in the twentieth century. Jakob Vogel, "Von der Wissenschafts- zur Wissensgeschichte: Für eine Historiesierung der Wissensgesellschaft," *Geschichte und Gesellschaft* 30, no. 4 (2004):639–60; 644. See also Scribner, "Reformation, popular magic, and the 'disenchantment of the world.'"

7. Max Weber, *Kapitalismens uppkomst: Urval och förord av Hans L. Zetterberg*, trans. Leif Björk (Göteborg: Timbro, 1986), 30, 39–41.

8. For an overview of positions, Dym, *Divining science*, 168–69.

Bibliography

MANUSCRIPT SOURCES

Kungliga Biblioteket (KB)
National Library of Sweden, Stockholm
Rinmanska samlingen. Apprentice contracts signed by Sven Rinman.
X277 Prober-konsten sammanskrefwen af Jacob Fischer proberare uti Kongl. Bärgs Collegio. Manuscript of Jacob Fischer's course in assaying signed by and written in the hand of Lars Schultze.
X273 Bergmästaren Erich Odhelii berättelse till Kongl. Maj:t om en resa till Tyskland, Ungern, Tirol, Biscaien, Engelland och Holland för vinnande af kännedom om Bergvärkens och metallernas der i provincierna tillstånd och afförsel. Erich Odhelius's journey to Germany, Hungary, Tyrol, Spain, England, and Holland.

Kungliga Svenska Vetenskapsakademiens Arkiv (KVA)
Archives of the Royal Swedish Academy of Sciences, Stockholm
Bergianska Brevsamlingen, vol. 13. Letter from Torbern Bergman to Bengt Bergius.
E1:1, F1:2, F1:3 Axel Fredrik Cronstedts arkiv. Axel Fredrik Cronstedt's manuscripts, notebooks, and letters.
E1:5, E1:7 Wargentins arkiv. Archive of Pehr Wilhelm Wargentin, various letters.

Riksarkivet (RA)
Swedish National Archives, Stockholm
A 1:35, D 2:15, D 6:3, D 6:6, D 6:9–D 6:13, E4. Bergskollegiums arkiv, huvudarkivet. Main archives of the Bureau of Mines.
1592 b Riksdagen 1746–47, Bergsdeputationen, vol. 2. Protocols of the Swedish Parliament, 1746–47, deputation for mining affairs, vol. 2.

Universiteitsbibliotheek Leiden
Volumina Inscriptorum [shelf mark ASF 13]

Uppsala Universitetsbibliotek (UUB)
Uppsala University Library, Uppsala
D 14 Urban Hiärnes manuskript. Urban Hiärne's manuscripts.
D 1433 [Anton von Swab, copy in the handwriting of Lars Schultze] Kort betänkande huru en yngling som täncker söka sin fortkomst vid Bergsväsende bör sin tid anlägga. Instructions for a young man intent on a career in mining.
D 1450a Föreläsningar i kemi. Lecture notes from Georg Brandt's chemistry course.
D 1459 d Manuscripta T. Bergman, vol. 4. (Biografiska och litteraturhistoriska anteckningar om kemister m. m.) Torbern Bergman's manuscript notes, draft biographies of chemists.
G21 Svensk brevväxling till Torbern Bergman. Letters to Torbern Bergman from Swedish correspondents.
X 240 Meritförteckning 1757. Georg Brandt's curriculum vita of 1757.
X241 Herr assessoren och Riddaren Rinmans lefverne och meriter. Sven Rinman's autobiography.
X268 Curriculum Vitae Meae. Urban Hiärne's autobiography.

PRIMARY PUBLISHED SOURCES
[Adlersparre, C.], ed. *Historiska samlingar.* Vol. 5. Stockholm: F. B. Nestius, 1822.
Agricola, Georgius. *De re metallica: Translated from the first Latin edition of 1556 ... by Herbert Clark Hoover ... and Lou Henry Hoover.* 1556; New York: Dover Publications, 1950.
———. "Die Lebewesen unter Tage." In Georgius Agricola, *Vermischte Schriften.* Vol. 1. Ed. Hans Prescher and trans. Georg Fraustadt. Berlin: VEB Deutcher Verlag der Wissenschaften, 1961.
Becher, Joh. Joachim. *Närrische Weißheit Und Weise Narrheit: Oder Ein Hundert so Politische als Physicalische / Mechanische und Mercantilische Concepten und Propositionen / deren etliche gut gethan/ etliche zu nichts worden.* 1707.
Bergman, Torbern. *Åminnelsetal öfver ... Anton von Swab ... Hållit i Stora Riddarhus-Salen den 29 Junii 1768.* Stockholm: Lars Salvius, 1768.
———. *Åminnelsetal öfver framledne Bergs-rådet och medicinae doctoren, samt K. Vetenskaps sällskapets i Upsala, och Kongl. Academiens i Stockholm ledamot, Herr Georg Brandt.* Stockholm: Lars Salvius, 1769.
———. *Anledning til föreläsningar öfver chemiens beskaffenhet och nytta, samt naturliga kroppars almännaste skiljaktigheter.* Stockholm: Uppsala och Åbo, 1779.

————. *Physical and chemical essays*. Trans. from the original Latin by Edmund Cullen. To which are added notes and illustrations, by the translator. Vol. 1. London: J. Murray, 1784.

————. *Physisk beskrifning öfver jord-klotet: på cosmographiska sällskapets vägnar författad*. Uppsala, 1766.

————. *Sciagraphia regni mineralis, secundum principia proxima digesti*. Leipzig, 1782.

————. *Torbern Bergman's foreign correspondence*. Ed. Göt Carlid and Johan Nordström. Uppsala: Almqvist och Wiksell, 1965.

Berndtson, Bernhard. *Tal, om en bergsmans kunskaper och gjöromål hållet för Kongl. Vetenskaps Academien vid praesidii nedläggande den 1 februar. 1797*. Stockholm: Joh. P. Lindh, 1797.

Boerhaave, Herman. *A new method of chemistry: Including the history, theory, and practice of the art*. Trans. Peter Shaw. 2nd ed. London, 1741.

Boyle, Robert. "The sceptical chymist." In Thomas Birch, ed., *Robert Boyle: The works*, 1:474–586. Hildesheim: Georg Olms, 1965.

Brandt, Georg. "De arsenico observations." *Acta literaria et scientiarum Sveciae* (1733): 39–43.

————. "Dissertatio de semi-metallis." *Acta literaria et scientiarum Sveciae* (1735): 1–12.

————. "Experimentum quo probatur dari attractionem mercurii in aurum." *Acta literaria et scientiarum Sveciae* (1731): 1–8.

————. *En grundelig anledning til mathesin universalem och algebram, efter herr And. Gabr. Duhres håldne praelectioner sammanskriven af Georg Brandt*. Stockholm: Joh. L. Horrn, 1718.

————. *Tal om färg-cobolter, hållit för Kongl. Vet. Acad. Vid praesidii nedläggande den 30 jul. 1760*. Stockholm: Lars Salvius, 1760.

Bromell, Magnus von. *Lithographiae Svecanae: Specimen primum, Calculos humanos, variaqve animalium concreta lapidea exhibens, Juxta seriem atque ordinem, quo in Musaeo Metallico Bromeliano servantur*. Uppsala, 1726.

————. *Lithographiae Svecanae: Specimen secundum, Telluris svecanae petrifacata lapidesque figuratos varios exhibens, Juxta seriem atque ordinem, quo in Museo Metallico Bromeliano servantur*. Uppsala, 1727.

————. *Mineralogia: Eller inledning til nödig kundskap at igenkiänna och upfinna allahanda berg-arter, mineralier, metaller samt fossilier*. 2nd ed. Stockholm: Gottfried Kiesewetter, 1739.

Bure, Johan [Johannes Bureus]. "Anteckningar af Johannes Thomae Agrivillensis Bureus." *Samlaren: Tidskrift utgiven af Svenska literatursällskapets arbetsutskott* 4(1883): 12–43 and 71–126.

Conway, Anne. *The principles of the most ancient and modern philosophy*. London, 1692.

Croll, Oswald. *Basilica chymica: Oder Alchymistisch Königlich Kleÿnod: Ein Philosophisch durch sein selbst eigene erfahrung confirmierte und bestättigte Beschreibüng und gebrauch der aller fürtrefflichen Chimischen Artzneÿen so auß dem Liecht der Gnaden und Natur genommen in sich begreiffen*. Frankfurt: Gottfried Tampachen, [1629].

Cronstedt, Axel Fredrik. *Åminnelsetal Öfver . . . Henric Teophil Scheffer . . . Hållit i Stora Riddarehus-Salen, den 17. September 1760.* Stockholm: Lars Salvius, 1760.

————. *Axel von Kronstedts Versuch einer Mineralogie: Aus neue as dem Schwedischen übersetzt und nächst verschiedenen Anmerkungen vorzüglich mit äussern Beschreibungen der Fossilien vermehrt von Abraham Gottlob Werner.* Leipzig: Siegfried Lebrecht Crusius, 1780.

————. *Cronstedts Versuch einer Mineralogie: Vermehret durch Brünnich.* Copenhagen and Leipzig: Prost und Rothens Erben, 1770.

————. *An essay towards a system of mineralogy: Translated from the original Swedish, with notes by Gustav von Engestrom. To which is added, a treatise on the pocket-laboratory, containing an easy method, used by the author, for trying mineral bodies, written by the translator. The whole revised and corrected, with some additional notes by Emanuel Mendes Da Costa.* London, 1770.

[————]. *Försök til Mineralogie, eller mineral-rikets upställning.* Stockholm, 1758.

————. "Fortsättning af rön och försök, gjorde med en malm-art från Los kobolt-grufvor." *Kungliga Vetenskapsakademiens Handlingar* (1754).

————. *Inträdes-tal om medel till mineralogiens vidare förkofran: Hållit för Kongl. Svenska Vetenskapsakademien den 9 Feb. 1754.* Stockholm: Lars Salvius, 1754.

————. "Några rön och anmärkningar vid Platina di Pinto." *Kungliga Vetenskapsakademiens Handlingar* (1764).

————. *Rön och anmärkningar vid Jämtlands mineral historia.* Facsimile reprint from *Kongl. Vetenskaps Academiens Handlingar,* 1763. Stockholm, 1993.

————. "Rön och försök gjorde med en malm-art, från Los kobolt grufvor i Färila Socken och Helsingeland." *Kungliga Vetenskapsakademiens Handlingar* (1751).

Ehrenheim, Fr. Wilh., ed. *Tessin och Tessiniana.* Stockholm: Johan Imnelius, 1819.

Ekström, Daniel. *Tal, om järn-förädlingens nytta och vårdande; hållit i Kongl. Vetenskapsacademien vid praesidii afläggning d. 28. apr. 1750.* Stockholm, 1750.

Engeström, Gustav von. "Tal, om vissa svårigheter och andra omständigheter, som möta vid utöfvandet af chymien; Hållet för Kongl. Vetensk. Academien." Stockholm, 1782.

Espagnet, Jean d'. *Enchyridion physicae restitutae; or, The summary of physicks recovered. Wherein the true harmonie of nature is explained, and many errours of the ancient philosophers, by canons and certain demonstrations, are clearly evidenced and evinced. / Written in Latine by the Duke of Espernon; and translated into English by Dr. Everard.* London: W. Bentley, [1651].

Garney, Johan Carl. *Handledning uti svenska masmästeriet.* Vols. 1–2. Stockholm, 1791.

Gjörwell, Carl Christoffer. *Det swenska biblioteket.* Vols. 1–2. Stockholm: Wildiska tryckeriet, 1757–58.

Hiärne, Urban. *Actorum Laboratorii Stockholmiensis Parasceve. Eller Förberedelse Til de undersökningar som uthi Kongl. Laboratorio äro genomgångne samt underrättelse om de förnämbste Grundstycken uthi Chymien.* Stockholm: Michaele Laurelio, 1706.

———. *Den besvarade och förklarade anledningens andra flock: Om jorden och landskap i gemeen.* Stockholm: Mich. Laurelio, 1706.

———. *Defensionis Paracelsicae Prodromus: Eller kort föremäle af then uthförligare förswars skrift för den stora philosophus theutonicus Theophrastus Paracelsus, som nyligen medelst en hård beskyllning uthan någon orsak är worden antastad.* Stockholm: Julius G. Matthiae, 1709.

———. *En ganska liten bergslykta: Förmedelst hvilken man sig uti den mörka bergsverkshandtering själf leda och lysa kan.* Ed. Carl Sahlin. Örebro: Örebro Dagblads tryckeri, 1909.

———. "En kort Beskrifning af min Resa från Stockholm genom Uppland, Gestricke- och Helsingeland til Herjedalen, Norige och Jemtland, dädan til Medelpad och Ångermanland, sedan genom Helsingeland samma vägen hem igen." In Carl Christoffer Gjörwell, *Nya svenska biblioteket,* 1:2. 1685; Stockholm: Peter Hessleberg, 1762.

———. *Den korta anledningen / til Åthskillige malm och bergarters / mineraliers och jordeslags efterspörjande och angifwande / beswarad och förklarad / jämte deras natur / födelse och i jorden tilwerckande / samt uplösning och anatomie i giörligaste mått skrifwen.* Stockholm, 1702.

———. Letter to Olof Hermelin, 24 April 1707. *Historisk Tidskrift* (1882): 264–68.

———. *Märkvärdigheter hos sjön Vettern.* Trans. Carl Stubbe. Stockholm: Norstedts, 1916. Originally published as "Memorabilia nonnulla lacus Vetteri," *Philosophical Transactions* 24 (1704–5), no. 298. London: S. Smith and B. Walford, 1705.

———. "Salig Landzhöfding Urban Hiernes Lefwernesbeskrifning af honom sielf i pennan fattad." In *Äldre svenska biografier,* ed. H. Schück, 5:139–79. Uppsala Universitets Årsskrift 1916 program 4:2. Uppsala: Akademiska Bokhandeln, 1916.

Jung-Stilling, Johann Heinrich. *Andelära framställd uti et med naturen, förnuftet och uppenbarelsen enligt swar på den frågan: Hvad bör man tro och icke tro om aningar, syner och andars uppenbarelser.* Swedish trans. from German. Göteborg, 1812.

Justi, Johann Heinrich Gottlob von. *Grundriss des gesamten Mineralreiches; worinnen alle Fossilien in einem, ihren wesentlichen Beschaffenheiten gemässen, Zusammenhange vorgestellet und beschrieben werden.* Göttingen, 1757.

Kant, Immanuel. *En andeskådares drömmar: I ljuset av metafysikens drömmar.* Swedish trans. from German with an introduction by Efraim Briem. Lund: Gleerup, 1921.

Kongl. stadgar, förordningar, bref och resolutioner, angående justitien och hushållningen wid bergwerken och bruken, första fortsättningen. Från och med år 1736 til och med år 1756. Vol. 2. Stockholm, 1786.

Kongl. stadgar, förordningar, bref och resolutioner, angående justitien och hushållningen wid bergwerken och bruken, andra fortsättningen. Ifrån och med år 1757 til och med år 1791. Vol. 3. Stockholm, 1797.

Leyonmarck, Gustav Adolph. *Tal, om utsigten för svenska bergshandteringen i framtiden; hållet, i Kongl. Maj:ts närvaro, för dess Vetenskaps Academie, vid praesidii nedläggande derstädes, den 26 april 1775.* Stockholm, 1775.

Lillienberg, Jean Georg. *Tal, om svenska bergshandteringens förmåner och hinder, hållet uti Hans Kongl. Maj:ts höga närvaro, för Kongl. Vetenskaps-Academien, vid praesidii nedläggande, den 7 febr. 1776.* Stockholm, 1776.

[Loenbom, Samuel], ed. *Handlingar til Konung Carl XI:tes Historia.* Pts. 12–13. Stockholm: Lars Salvius, 1772.

Magnus, Olaus. *Historia om de nordiska folken: Deras olika förhållanden och villkor, plägseder, religiösa och vidskepliga bruk, färdigheter och idrotter, samhällsskick och lefnadssätt, krig, byggnader, och redskap, grufvor och bergverk, underbara ting samt om nästan alla djur, som lefva i Norden och deras natur.* Vol. 2, pts. 1–5. 1555; Uppsala and Stockholm: Almqvist och Wiksell, 1909–51.

Naudaeus, Gabriel. *The history of magick by way of apology, for all the wise men who have unjustly been reputed magicians, from the creation, to the present age. Written in French, by G. Naudaeus Late Library-Keeper to Cardinal Mazarin.* Trans. J. Davies. 1625; London: John Streater, 1657.

Paracelsus. *Theophrast von Hohenheim gen. Paracelsus, Sämtliche Werke 1. Abteilung Medizinische naturwissenschaftliche und philosophische Schriften.* Ed. Karl Sudhoff. Pts. 3, 13, 14. Munich and Berlin, 1930–31.

Polhem, Christopher. *Christopher Polhems efterlämnade skrifter III: Naturfilosofiska och fysikaliska skrifter redigerade av Axel Liljecrantz.* Uppsala: Almqvist och Wiksell, 1952–53.

Pott, Johann Heinrich. *Fortsetzung dererer Chymischen Untersuchungen, welche von der Lithogeognosie oder Erkäntniss und Bearbeitung derer Steine und Erden specieller handeln.* Berlin and Potsdam: Christian Friedrich Voss, 1751.

———. *Send-Schreiben an den Herrn Berg-Rath von Justi darin die Einwürfe, die er ihm in seinen wieder aufgelegten Chymischen Schriften von neuen gemacht hat, erörtert und abgelehnet, und die darinn angefochtene Chymisch-Physicalische Materien weiter untersucht und ausgeführt werden.* Berlin: Spenerischen Buchhandlung, 1760.

Quist, Bengt. "Rön om bly-erts." *Kungliga Vetenskapsakademiens Handlingar* (1754).

Rinman, Sven. *Allgemeines Bergwerkslexikon: Nach dem Schwedischen Original bearbeitet und nach den neusten Entdeckungen vermehrt von einer Gesellschaft deutschr Gelehrten und Mineralogen.* 2 vols. Leipzig, 1808.

———. *Åminnelsetal öfver framledne Bergmästaren . . . Cronstedt.* Stockholm, 1766.

———. *Bergwerks lexicon.* 2 vols. Stockholm: Johan A. Carlbohm, 1788–89.

———. *Inledningar til kunskap om den gröfre jern- och stål-förädlingen och des förbättrande.* Stockholm, 1772.

Rosen, Carl von, ed. *Bref från Samuel Bark till Olof Hermelin 1702–1708: Senare delen 1705–1708.* Stockholm: Norstedt, 1915.

Scheffer, Henric Theophil. "Det hvita gullet, eller sjunde metallen. Kalladt i Spanien Platina del Pinto, Pintos små silfver, beskrifvit til sin natur." *Kungliga Vetenskapsakademiens Handlingar* (1752): 269–75.

———. "Tilläggning om samma metal." *Kungliga Vetenskapsakademiens Handlingar* (1752): 276–78.

Schröderstierna, Samuel. *Berättelser över de finare järn- stål- och metallfabrikerna i Sverige åren 1754–1759*. Pts. 1–2. Issued by Gösta Malmborg with an introduction by Carl Sahlin. Stockholm: Generalstabens litografiska anstalt, 1925.

———. *Bergsmannanäringens nytta och skötsel, förestäld uti et tal til Kongl. Svenska Vetenskapsakademien den 28 october 1749*. Stockholm, 1749.

Stockenström, Erik von. *Bergsmannanäringens nytta och skötsel, förestäld uti et tal til Kongl. Svenska Vetenskapsakademien den 28 october 1749*. Stockholm, 1749.

Swedenborg, Emanuel. *Om järnet och de i Europa vanligaste vedertagna järnframställningssätten . . . under redaktion av Hj. Sjögren*. 1734; Stockholm: Wahlström och Widstrand, 1923.

Swift, Jonathan. *Gulliver's travels*. 1726.

Tafel, R. L., ed. *Documents concerning the life and character of Emanuel Swedenborg: Collected, translated, and annotated by R. L. Tafel*. Vols. 1–3. London: Swedenborg Society, 1875–77.

Tilas, Daniel. *Åminnelse-tal öfver . . . Gustav Bonde . . . Hållit i Stora Riddarehus-Salen Den 7. Junii 1766*. Stockholm: Lars Salvius, 1766.

———. "Anteckningar ur Hiärne Ättens Minne." In [C. Adlersparre], ed., *Historiska samlingar*, 5:432–42. Stockholm: F. B. Nestius, 1822.

———. *En Bergsmans rön och försök i mineralriket*. Åbo, 1738.

———. "Curruculum vitae 1–2, 1712–1757 samt fragment av dagbok september—oktober 1767." In *Historiska Handlingar*, 38:1. Stockholm: Norstedts, 1966.

Trofast, Jan, ed. *Johan Gottlieb Gahn Brev: Utgivna med kommentarer av Jan Trofast*. Pt. 2. Lund: Jan Trofast, 1994.

Trolle-Bonde Carl. *Riksrådet grefve Gustav Bonde 2: Gustaf Bondes Litterära verksamhet*. Anteckningar om Bondesläkten. Lund: Berlingska, 1897.

Wallerius, Johan Gottschalk. *Bref om chemiens rätta beskaffenhet nytta och wärde*. Stockholm, 1751.

———. *Mineralogia eller mineralriket indelt och beskrifwit*. Stockholm, 1747.

SECONDARY PUBLISHED SOURCES

Abbri, Fernando. "Some ingenious systems: Lavoisier and the northern chemistry." In M. Beretta, ed., *Lavoisier in perspective*, 95–108. Deutsches Museum Abhandlungen und Berichte, Neue Folge, Band 21. Munich: Deutsches Museum, 2005.

Åkerman, Susanna. *Fenixelden: Drottning Kristina som alkemist*. Gidlunds: Möklinta, 2013.

Almgren, Oscar, Isak Collijn, Gunnar Collin, Axel Nilsson, and Rutger Sernander. "Utgifvarnas förord." In Olaus Magnus, *Historia om de nordiska folken: Deras olika förhållanden och villkor, plägseder, religiösa och vidskepliga bruk, färdigheter och idrotter, samhällsskick och lefnadssätt, krig, byggnader, och redskap, grufvor och bergverk, underbara ting samt om nästan alla djur, som lefva i Norden och deras natur*, 1:iii–v. 1555; Uppsala: Almqvist och Wiksell, 1909.

Almquist, Johan Axel. *Bergskollegium och Bergslagsstaterna 1637–1857: Administrativa och biografiska anteckningar*. Meddelanden från Svenska Riksarkivet 2:3. Stockholm, 1909.

Ankarloo, Bengt. *Trolldomsprocesserna i Sverige*. 2nd ed. Rättshistoriskt bibliotek 17. Stockholm: Nordiska Bokhandeln, 1984.

Askegren, Katrin. "Provinsen Ingermanland under Hiärnes tid–en kort historik." In S. Ö. Ohlson and S. Tomingas-Joandi, eds., *Den otidsenlige Urban Hiärne: Föredrag från det internationella Hiärne-symposiet i Saadjärve, 31 augusti–4 september 2005*, 35–43. Tartu: Trükikoda Greif, 2008.

Asker, Björn. "Paykull, Otto Arnold." In *Svenskt biografiskt lexikon*, 28:768–70.

Asplund, Ingemark Camilla. *The genre of trolls: The case of a Finland-Swedish folk belief tradition*. Åbo: Åbo Akademi University Press, 2004.

Bartels, Christoph. "Der Bergbau vor der hochindustriellen Zeit—Ein Überblick." In R. Slotta and C. Bartels, eds., *Meisterwerke bergbaulicher Kunst vom 13. bis 19. Jahrhundert*. Bochum: Deutchen Bergbau-Museum, 1990.

———. "The production of silver, copper, and lead in the Harz Mountains from late medieval times to the onset of industrialization." In U. Klein and E. Spary, eds., *Materials and expertise in early modern Europe: Between market and laboratory*, 71–100. Chicago and London: University of Chicago Press, 2010.

———. *Vom früneuzeitliche Montangewerbe zur Bergbauindustrie: Erzgebau im Oberharz 1635–1866*. Bochum: Deutsches Bergbau-Museum, 1992.

Baumgärtel, Hans. *Bergbau und Absolutismus: Der sächsische Bergbau in der zweiten Hälfte des 18. Jahrhunderts und Massnamen zu seiner Verbesserung nach dem Siebenjährigen Kriege*. Freiberger Forschungshefte D 44 Kultur on Technik. Leipzig: VEB Deutscher Verlag für Grundstoffindustrie, 1963.

Berg, Bjørn Ivar. "Das Bergseminar in Kongsberg in Norwegen, 1757–1814." *Der Anschnitt: Zeitschrift für Kunst und Kultur im Bergbau* 3–4(2008): 152–65.

———. *Gruveteknikk ved Kongsberg Sølvverk 1623–1914*. Dragvoll: Norges teknisk-naturvitenskapelige universitet, 1998.

Bergquist, Lars. "Kusinfejden: Swedenborg kontra Linné." In E. Pierre and A. W. Johansson, eds., *Över vida fält: Studier i utrikespolitik, diplomati och historia Vänbok till Mats Bergquist*, 37–45. Lund: Sekel, 2008.

———. *Swedenborgs hemlighet: Om Ordets betydelse, änglarnas liv och tjänsten hos Gud*. Stockholm: Natur och Kultur, 1999.

Bloor, David. *Knowledge and social imagery*. 2nd ed. 1976; Chicago: University of Chicago Press, 1991.

Bostridge, Ian. *Witchcraft and its transformations, c. 1650–c. 1750*. Oxford: Clarendon Press, 1997.

Brianta, Donata. "Education and training in the mining industry, 1750–1860: European models and the Italian case." *Annals of Science* 57(2000): 268–75.

Bring, Samuel. "Bidrag till Christopher Polhems lefnadsteckning." In S. Bring, ed., *Christopher Polhem: Minnesskrift utgifven af Svenska Teknologiföreningen*, 3–119. Stockholm: Centraltryckeriet, 1911.

Buchanan, Brenda J. "The art and mystery of making gunpowder: The English experience in the seventeenth and eighteenth centuries." In B. D. Steele and T. Dorland, eds., *The heirs of Archimedes: Science and the art of war through the age of Enlightenment*, 235–74. Cambridge: MIT Press, 2005.

Burns, Robert M. *The great debate on miracles: From Joseph Glanvill to David Hume*. Lewisburg, PA: Bucknell University Press, 1981.

Cajdert, Mats Ola. "En Hjärne-lärjunge i Europa: Erik Odelstiernas Brev till Urban Hjärne 1683–1687." *Personhistorisk Tidskrift* 92 (1996): 1–23.

Carlquist, G. "Gustav Bonde." In *Svenskt biografiskt lexikon*, 5:362 –77.

Cassebaum, Heinz, and George B. Kauffman. "The analytical concept of a chemical element in the work of Bergman and Scheele." *Annals of Science* 33 (1976): 447–56.

Cavallin, Maria. *I kungens och folkets tjänst: Synen på den svenske ämbetsmannen 1750–1780*. Göteborg: Historiska institutionen, 2003.

Chaiklin, Martha. "Simian amphibians: The mermaid trade in early modern Japan." In Yoko Nagazumi, ed., *Large and broad: The Dutch impact on early modern Asia: Essays in honor of Leonard Blussé*, 241–73. Tokyo: Toyo Bunko, 2010.

Clark, Stuart. *Thinking with demons: The idea of witchcraft in early modern Europe*. Oxford: Clarendon Press, 1997.

Cole, William A. *Chemical literature, 1700–1860: A bibliography with annotations, detailed descriptions, comparisons and locations*. London: Mansell, 1990.

Cook, Harold J. *Matters of exchange: Commerce, medicine, and science in the Dutch golden age*. New Haven and London: Yale University Press, 2007.

Copenhaver, Brian. "Natural magic, hermetism, and occultism in early modern science." In D. C. Lindberg and R. S. Westman, eds., *Reappraisals of the scientific revolution*, 261–301. Cambridge: Cambridge University Press, 1990.

Daston, Lorraine, ed. *Biographies of scientific objects*. Chicago and London: University of Chicago Press, 2000.

Daston, Lorraine, and Katharine Park. *Wonders and the order of nature: 1150–1750*. New York: Zone Books, 1998.

Day, Joan, and R. F. Tylecote. *The industrial revolution in metals*. London: Institute of Metals, 1991.

Debus, Allen G., "Fire analysis and the elements in the sixteenth and seventeenth centuries." *Annals of Science* 23 (1967): 127–47.

Dobbs, Betty Jo Teeter. *The foundations of Newton's alchemy, or, "The hunting of the greene lyon."* Cambridge: Cambridge University Press, 1975.

Dolan, Brian. "Transferring skill: Blowpipe analysis in Sweden and England, 1750–1850." In B. Dolan, ed., *Science unbound: Geography, space and discipline*, 91–125. Umeå, 1998.

Dunér, David. "Naturens alfabet: Polhem och Linné om växternas systematik." *Svenska linnésällskapets årsskrift* (2008): 31–70.

———. *Världsmaskinen: Emanuel Swedenborgs naturfilosofi*. Lund and Nora: Nya Doxa, 2004.

Dym, Warren. *Divining science: Treasure hunting and earth science in early modern Germany*. Leiden: Brill, 2011.

Eddy, Matthew D. *The language of mineralogy: John Walker, chemistry and the Edinburgh Medical School, 1750–1800*. Farnham, UK, and Burlington, VT: Ashgate, 2008.

Edenborg, Carl Michael. *Gull och mull*. Lund: Ellerströms, 1997.

Eriksson, Gunnar. *Rudbeck 1630–1702: Liv, lärdom, dröm i barockens Sverige.* Stockholm: Atlantis, 2002.

Erixon, Sigurd. "Some examples of popular conceptions of sprites and other elementals in Sweden during the 19th century." In Å. Hultkrantz, ed., *The supernatural owners of nature: Nordic symposium on the religious conceptions of ruling spirits. genii loci, genii speciei and allied concepts*, 34–37. Stockholm and Uppsala: Almqvist and Wiksell, 1961.

Evans, Chris, and Göran Rydén. *Baltic iron in the Atlantic world in the eighteenth century.* Leiden and Boston: Brill, 2007.

Fessner, Michael. "Die Berliner Bergakademie von ihrer Gründung bis zur Eingliderung in die Technische Hochschule Berlin-Charlottenburg. 1770–1916." In A. Westermann and E. Westermann, eds., *Wirtschaftslenkende Montanverwaltung- Fürstlicher Unternehmer—Merkantilismus: Zusammenhänge zwischen der Ausbildung einer fachkompetenten Beamtenschaft in der staatlichen Geld- und Wirtschaftspolitik in der Frühen Neuzeit*, 439–62. Husum: Matthiesen Verlag, 2009.

Findlen, Paula. "Inventing nature: Commerce, art and science in the early modern cabinet of curiosities." In P. H. Smith and P. Findlen, *Merchants and marvels: Commerce, science, and art in early modern Europe*, 297–323. New York and London: Routledge, 2002.

Florschütz, Gottlieb. *Swedenborgs verborgene Wirkung auf Kant: Swedenborg und die okkulten Phänomene aus der Sicht von Kant und Schopenhauer.* Epistemata, Würzburger Wissenschaftliche Schriften, Reihe Philosophie, Band 106. Würzburg: Königshausen und Neumann, 1992.

Fors, Hjalmar. "'Away, away to Falun!' J. G. Gahn and the application of Enlightenment chemistry to smelting." *Technology and Culture* (2009): 549–68.

———. "J. G. Wallerius and the laboratory of Enlightenment." In E. Baraldi, H. Fors, and A. Houltz, eds., *Taking place: The spatial contexts of science, technology and business*, 3–33. Sagamore Beach: Science History Publications, 2006.

———. "Kemi, paracelsism och mekanisk filosofi: Bergskollegium och Uppsala cirka 1680–1770." *Lychnos* (2007): 211–44.

———. "Matematiker mot linneaner: Konkurrerande vetenskapliga nätverk kring Torbern Bergman." In S. Widmalm, ed., *Vetenskapens sociala strukturer: Sju historiska fallstudier om konflikt, samverkan och makt*, 25–53. Lund: Nordic Academic Press, 2008.

———. *Mutual favours: The social and scientific practice of eighteenth-century Swedish chemistry.* Uppsala: Uppsala Universitet, 2003.

———. "Occult traditions and enlightened science: The Swedish Board of Mines as an intellectual environment, 1680–1760." In L. Principe, ed., *Chymists and chymistry: Studies in the history of alchemy and early modern chemistry*, 239–52. Sagamore Beach: Science History Publications, 2007.

———. "Speaking about the other ones: Swedish chemists on alchemy, c. 1730–70." In J. R. Bertomeu-Sánchez, D. T. Burns, and B. Van Tiggelen, eds., *Neighbours and territories: The evolving identity of chemistry proceedings of the 6th International Conference on the History of Chemistry*, 283–89. Leuven, 2008.

———. "Stockenström, Erik von." In *Svenskt biografiskt lexikon*, 164:548–53.

———. "Vetenskap i alkemins gränsland: Om J. G. Wallerius *Wattu-riket*." *Svenska Linnésällskapets Årsskrift* (1996–97): 33–60.

Frängsmyr, Tore. "The Enlightenment in Sweden." In R. Porter and M. Teich, eds., *The Enlightenment in national context*, 164–75. Cambridge: Cambridge University Press, 1981.

———. *Geologi och skapelsetro: Föreställningar om jordens historia från Hiärne till Bergman*. Uppsala, 1969.

———. *Sökandet efter upplysningen: perspektiv på svenskt 1700-tal*. Höganäs: Wiken, 1993.

Fritz, Martin, ed. *Iron and steel on the European market in the seventeenth century: A contemporary Swedish account of production forms and marketing*. Stockholm: Berlings, 1982.

Gadelius, Bror. *Urban Hjärne och häxprocesserna i Stockholm 1676*. Offprint from *Hygiea*, 1909. Stockholm: Isaac Marcus, 1909.

Galán, A., and R. Moreno. "Platinum in the eighteenth century: A further Spanish contribution to an understanding of its discovery and early metallurgy." *Platinum Metals Review* 36 (1992): 40–47.

Gerentz, Sven. *Kommerskollegium och näringslivet: Minnesskrift utarbetad av Sven Gerentz på uppdrag av Kungl. Kommerskollegium till erinran om Kollegii 300-åriga ämbetsförvaltning 1651–1951*. Stockholm: Kommerskollegium, 1951.

Gieryn, Thomas F. *Cultural boundaries of science: Credibility on the line*. Chicago and London: University of Chicago Press, 1999.

Gillingstam, Hans. "Hermelin, släkt." In *Svenskt biografiskt lexikon*, 18:702.

Gleeson, Janet. *The Arcanum: The extraordinary true story*. New York: Warner, 1998.

Golinski, Jan. "Chemistry." In R. Porter, ed., *The Cambridge history of science*, vol. 4: *Eighteenth-century science*. Cambridge: Cambridge University Press, 2003.

Grage, Elsa-Britta. "Odelstierna, Erich." In *Svenskt biografiskt lexikon*, 28:34–37.

Greenaway, Frank. "Pott, Johann Heinrich." In C. C. Gillispie, ed., *Dictionary of scientific biography*, 11:109.

Guerlac, Henry. "Some French antecedents of the chemical revolution." *Chymia: Annual Studies in the History of Chemistry* 5 (1959).

Hacking, Ian. *Historical ontology*. Cambridge: Harvard University Press, 2004.

Heckscher, Eli. *Sveriges ekonomiska historia från Gustav Vasa: Före frihetstiden*. Pt. 1:2. Stockholm: Bonniers, 1936.

Heinrichs, Michael. *Emanuel Swedenborg in Deutschland: Eine kritische Darstellung der Rezeption des schwedischen Visionärs im 18. und 19. Jahrhundert*. Europäische Hochschulschriften: Reihe 20, Philosophie, Band 47. Frankfurt am Main: Peter D. Lang, 1979.

Herrmann, Walther. *Bergrat Henckel: Ein wegbereiter der Bergakademie*. Freiberger Forschungshefte D 37 Kultur on Technik. Berlin: Akademie-Verlag,1962.

Hildebrand, Bengt. *Kungl. Svenska vetenskaps akademien: Förhistoria, grundläggning och första organisation*. Stockholm: K. Vetenskapsakademien, 1939.

Hildebrand, Karl-Gustaf. "Gammalt och nytt i det svenska järnets historia: En översikt över fem årtionden." *Daedalus: Tekniska museets årsbok* 65 (1997): 1–30.

———. *Swedish iron in the seventeenth and eighteenth centuries: Export industry before the industrialization.* Stockholm: Jernkontoret, 1992.

Hoerder, Dirk. *Cultures in contact: World migrations in the second millennium.* Durham and London: Duke University Press, 2002.

Hofberg, Herman. *Svenskt biografiskt handlexikon: Alfabetiskt ordnade lefnadsteckningar af Sveriges namnkunniga män och kvinnor från reformationen till nuvarande tid* 2. Stockholm: Bonniers, 1906.

Holmes, Frederic L. "Chemistry." In A. C. Kors, ed. in chief, *Encyclopedia of the Enlightenment,* vol. 1. Oxford: Oxford University Press, 2003.

Hult, O. T. "Bromell, Magnus von." In *Svenskt biografiskt lexikon,* 6:392–401.

Hultkrantz, Åke, ed. *The supernatural owners of nature: Nordic symposion on the religious conceptions of ruling spirits. genii loci, genii speciei and allied concepts.* Stockholm and Uppsala: Almqvist and Wiksell, 1961.

Hunt, L. B. "Swedish contributions to the discovery of platinum: The researches of Scheffer and Bergman." *Platinum Metals Review* 24 (1980): 31–36.

Hunt, Lynn, and Margaret Jacob. "Enlightenment studies." In A. C. Kors, ed. in chief, *Encyclopedia of the Enlightenment,* 1:418–30. Oxford: Oxford University Press, 2003.

Hunter, Michael. "Introduction." In M. Hunter, ed., *The occult laboratory: Magic, science and second sight in late seventeenth-century Scotland. A new edition of Robert Kirk's The Secret Commonwealth and other texts, with an introductory essay by Michael Hunter,* 1–32. Woodbridge, UK: Boydell Press, 2001.

Häll, Jan. *I Swedenborgs labyrint: Studier i de gustavianska swedenborgarnas liv och tänkande.* Stockholm: Atlantis, 1995.

Häll, Mikael. *Skogsrået, näcken och djävulen: Erotiska naturväsen och demonisk sexualitet i 1600-och 1700-talets Sverige.* Stockholm: Malörts förlag, 2013.

Högberg, Staffan. "Inledning." In Anton von Swab, *Anton von Swabs berättelse om Avesta kronobruk 1723,* 7–30. Jernkontorets bergshistoriska skriftserie 19. Stockholm: Jernkontoret, 1983.

Isacson, Maths. "Bergskollegium och den tidigindustriella järnhanteringen." *Daedalus: Tekniska museets årsbok* 66 (1998): 43–58.

Israel, Jonathan. *Enlightenment contested: Philosophy, modernity and the emancipation of man, 1620–1752.* Oxford: Oxford University Press, 2006.

———. *Radical enlightenment: Philosophy and the making of modernity, 1650–1750.* Oxford: Oxford University Press, 2001.

Johannisson, Karin. "Naturvetenskap på reträtt: En diskussion om naturvetenskapens status under svenskt 1700-tal." *Lychnos* (1979–80).

Kaiserfeld, Thomas. *Krigets salt: Salpeterisjudning som politik och vetenskap i den svenska skattemilitära staten under frihetstid och gustaviansk tid.* Lund: Sekel, 2009.

Karlsson, Åsa. "Nyen." In *Karolinska förbundets årsbok 1999, Att illustrera stormakten: Den svenska Fortifikationens bilder 1654–1719,* 60–61. Lund: Historiska Media, 2001.

Karlsson, Thomas. *Götisk kabbala och runisk alkemi: Johannes Bureus och den götiska esoterismen.* Stockholm, 2010.

Keller, Alex, "The age of the projectors." *History Today* 16 (1966): 467–74.

Kim, Mi Gyung. "Lavoisier, the father of modern chemistry?" In M. Beretta, ed., *Lavoisier in perspective,* 167–91. Munich: Deutsches Museum, 2005.

Klein, Ursula. "The chemical workshop tradition and the experimental practice: Discontinuities within continuities." *Science in Context* 9 (1996).

———. "Experimental history and Herman Boerhaave's *Chemistry of plants.*" *Studies in History and Philosophy of Science Part C: Studies in History and Philosophy of Biological and Biomedical Sciences* 34 (2003).

———. "The Prussian mining official Alexander von Humboldt." *Annals of Science* 69, no. 1: 27–68.

Klein, Ursula, and Wolfgang Lefèvre. *Materials in eighteenth-century science: A historical ontology.* Cambridge and London: MIT Press, 2007.

Klein, Ursula, and E. C. Spary. "Introduction: Why materials?" In U. Klein and E. C. Spary, eds., *Materials and expertise in early modern Europe: Between market and laboratory,* 1–23. Chicago and London: University of Chicago Press, 2010.

———, eds. *Materials and expertise in early modern Europe: Between market and laboratory.* Chicago and London: University of Chicago Press, 2010.

Koerner, Lisbet. "Daedalus hyperboreus: Baltic natural history and mineralogy in the Enlightenment." In W. Clark, J. Golinski, and S. Schaffer, eds., *The sciences in Enlightened Europe.* Chicago and London: University of Chicago Press, 1999.

———. *Linnaeus: Nature and nation.* Cambridge and London: Harvard University Press, 1999.

Konečný, Peter. "The hybrid expert in the 'Bergstaat': Anton von Ruprecht as a professor of chemistry and mining and as a mining official, 1779–1814." *Annals of Science* 69, no. 3 (2012): 335–47.

Kors, Alan Charles. Preface. In A. C. Kors, ed. in chief, *Encyclopedia of the Enlightenment,* 1:xvii–xxii. Oxford: Oxford University Press. 2003.

Ladenburg, Albert. "Kunckel, Johann von." In *Allgemeine Deutsche Biographie,* 17:376–77. Historischen Kommission bei der Bayerischen Akademie der Wissenschaften, 1883.

Lagerqvist, Lars O., Ulf Nordlind, and Hans Hirsch. *Från malm till mynt och medalj: Svenskt guld och silver ur Julius Haganders samling.* Boda Kyrkby: Vincent Förlag, 1998.

Lamm, Martin. *Upplysningstidens romantik: Den mystiskt sentimentala strömningen i svensk litteratur.* Pt. 1. 1918; Lund: Geber, 1963.

Lappalainen, Mirkka. "Släkt och stånd i bergskollegium före reduktionstiden." *Historisk tidskrift för Finland* 87, no. 2 (2002): 145–72.

———. *Släkten—makten—staten: Creutzarna i Sverige och Finland under 1600-talet.* Trans. Ann-Christine Relander. Stockholm: Norstedts, 2007.

Laudan, Rachel. *From mineralogy to geology: The foundations of a science, 1650–1830.* Chicago and London: University of Chicago Press, 1987.

Legnér, Mattias. *Fäderneslandets rätta beskrivning: Mötet mellan antikvarisk forskning och ekonomisk nyttokult i 1700-talets Sverige.* Helsingfors/Helsinki: Svenska Litteratursällskapet i Finland, 2004.

Lenhammar, Harry. *Tolerans och bekännelsetvång: Studier i den svenska swedenborgianismen 1765–95*. Uppsala: Uppsala Universitet,1966.

Lindborg, Rolf. *Descartes i Uppsala: Striderna om 'Nya Filosofien' 1663–1689*. Stockholm: Almqvist och Wiksell, 1965.

———. "Urban Hiärne såsom studerande för medicinprofessorn Petrus Hoffwenius i Uppsala." In S. Ö. Ohlson and S. Tomingas-Joandi, eds., *Den otidsenlige Urban Hiärne: Föredrag från det internationella Hiärne-symposiet i Saadjärve, 31 augusti–4 september 2005*, 45–50. Tartu: Trükikoda Greif, 2008.

Lindeboom, G. A. "Boerhaave, Hermann." In *Dictionary of scientific biography*, 1–2:224–28.

Lindgren, Michael. "Polhem, Christopher." In *Svenskt biografiskt lexikon*, 29:388–93.

Lindquist, David. *Studier i den svenska andaktslitteraturen under stormaktstidevarvet: Med särskild hänsyn till bön-, tröste-, och nattvardsböcker*. Uppsala: Almqvist och Wiksell, 1939.

Lindqvist, Svante. *Technology on trial: The introduction of steam power technology into Sweden, 1715–1736*. Uppsala: Almqvist and Wiksell, 1984.

Lindroth, Sten. *Christopher Polhem och Stora Kopparberget: Ett bidrag till bergsmekanikens historia*. Uppsala: Almqvist och Wiksell, 1951.

———. *Gruvbrytning och kopparhantering vid Stora Kopparberget intill 1800-talets börja: Del 1 Gruvan och gruvbrytningen*. Uppsala: Almqvist och Wiksell, 1955.

———. "Hiärne, Block och Paracelsus: En redogörelse för paracelsusstriden 1708–1709." *Lychnos* (1941): 191–229.

———. "Hiärne, Urban." In *Svenskt biografiskt lexikon*, 19:141–50.

———. *Kungliga Svenska vetenskapsakademiens historia 1739–1818*. Vol. 1, pt. 1. Stockholm: Kungl. Vetenskapsakademien, 1967.

———. "Linné—Legend och verklighet." *Lychnos* (1965–66): 56–122.

———. "Naturvetenskaperna och kulturkampen under frihetstiden." *Lychnos* (1957): 181–93.

———. "De stora häxprocesserna." In J. Cornell, S. Carlsson, J. Rosén, and G. Grenholm, eds., *Den svenska historien 7: Karl X Gustav, Karl XI. Krig och reduktion*, 158–61. Stockholm: Bonniers, 1978.

———. *Svensk lärdomshistoria: Frihetstiden*. Stockholm: Norstedt, 1975.

———. "Urban Hiärne och Laboratorium Chymicum." *Lychnos* (1946–47).

Lönnqvist, Bo. "17. Troll och människor." In B. Lönnqvist, *De andra och det annorlunda: Etnologiska texter*, 149–57. Helsingfors: Ekenäs tryckeri, 1996.

Lotman, Piret. "Kyrkoherdens son från Nevastranden: Föräldrar och barndomsmiljö." In S. Ö. Ohlson and S. Tomingas-Joandi, eds., *Den otidsenlige Urban Hiärne: Föredrag från det internationella Hiärne-symposiet i Saadjärve, 31 augusti—4 september 2005*, 25–33. Tartu: Trükikoda Greif, 2008.

Lundgren, Anders. "Bergshantering och kemi i Sverige under 1700-talet." *Med Hammare och Fackla* 29 (1985): 90–124.

———. "Vetenskap som vardagspraktik: Artefakter och dagligt arbete i ett kemiskt laboratorium under 1700- och 1800-talen." In Sven Widmalm and Hjalmar Fors, eds., *Artefakter: Industrin, vetenskapen och de tekniska nätverken*, 189–216. Hedemora: Gidlund, 2004.

Magnusson, Lars. *Merkantilism: Ett ekonomiskt tänkande formuleras*. Kristianstad: SNS förlag, 1999.

Mansikka, Tomas. "'Helias cum veniet restitutet omnia': Paracelsism, alkemi och reformm i 1600-talets protestantism." Licenciate dissertation. Åbo: Åbo akademi, 1998.

Margócsy, Dániel. "A museum of wonders or a cemetery of corpses: The commercial exchange of anatomical collections in early modern Netherlands." In Sven Dupré and Christoph Lüthy, eds., *Silent messengers: The circulation of material objects of knowledge in the early modern Low Countries*, 185–215. Berlin: LIT Verlag, 2011.

Martinón-Torres, Marcos. "The tools of the chymist: Archaeological and scientific analyses of early modern laboratories." In L. M. Principe, ed., *Chymists and chymistry: Studies in the history of alchemy and early modern chemistry*, 149–63. Sagamore Beach: Science History Publications, 2007.

Mattsson, Annie. *Komediant och riksförrädare: Handskriftcirkulerade smädeskrifter mot Gustaf III*. Uppsala: Uppsala Universitet, 2010.

Meinel, Christoph. "Theory or practice? The eighteenth-century debate on the scientific status of chemistry." *Ambix* (1983).

Melhado, Evan. "Mineralogy and the autonomy of chemistry around 1800." *Lychnos* (1990).

Merchant, Carolyn. *The death of nature: Women, ecology and the scientific revolution*. San Francisco: Harper and Row, 1980.

Moran, Bruce T. "German prince-practitioners: Aspects in the development of courtly science, technology, and procedures in the Renaissance." *Technology and Culture* 22, no. 2 (1981): 253–74.

———. "Patronage and institutions: Courts, universities, and academies in Germany: An overview, 1550–1750." In B. T. Moran, ed., *Patronage and institutions: Science, technology and medicine at the European court, 1500–1750*. Woodbridge, UK: Boydell Press, 1991.

Newman, William R. *Gehennical fire: The lives of George Starkey, an American alchemist in the scientific revolution*. Cambridge: Harvard University Press, 1994.

———. "The significance of 'chymical atomism.'" In Edith Dudley Sylla and William R. Newman, eds., *Evidence and interpretation in studies on early science and medicine*, 248–64. Leiden and Boston: Brill, 2009.

———. "What have we learned from the recent historiography of alchemy?" *Isis* 102 (2011): 313–21.

Newman, William R., and Lawrence Principe. "Alchemy vs. chemistry: The etymological origins of a historiographic mistake." *Early Science and Medicine* 3, no. 1 (1998): 38–41.

Norberg, Petrus. *Sala gruvas historia under 1500- och 1600-talen*. Sala: Salapostens boktryckeri AB, 1978.

Nummedal, Tara. *Alchemy and authority in the Holy Roman Empire*. Chicago and London: University of Chicago Press, 2007.

Odén, Sven. "Brandt, Georg." In *Svenskt biografiskt lexikon*, 5:784–89.

Ohlson, S. Ö., and S. Tomingas-Joandi, eds. *Den otidsenlige Urban Hiärne: Föredrag från det internationella Hiärne-symposiet i Saadjärve, 31 augusti—4 september 2005*. Tartu: Trükikoda Greif, 2008.

Olsson, Daniels Sven. *Falun mine*. Falun: Stiftelsen Stora Kopparberget, 2010.

Oja, Linda. *Varken Gud eller natur: Synen på magi i 1600- och 1700-talets Sverige*. Stockholm/Stehag: Symposion, 1999.

Olsson, Hugo. *Kemiens historia i Sverige intill år 1800*. Uppsala: Lychnosbibliotek, 1971.

Orrje, Jacob. *A mechanical state: The mathematical sciences and public office in the eighteenth-century Sweden*. Forthcoming 2015 diss., Uppsala.

Outram, Dorinda. *The Enlightenment*. 1995; Cambridge: Cambridge University Press, 2005.

Partington, J. R. *A history of chemistry*. Vol. 3. London: Macmillan, 1962.

Pérez, Liliane. "Technology, curiosity and utility in France and England in the eighteenth century." In B. Bensaude-Vincent and C. Blondel, eds., *Science and spectacle in the European Enlightenment*, 25–42. Aldershot, UK: Ashgate, 2008.

Persson, Mathias. *Det nära främmande: Svensk lärdom och politik i en tysk tidning, 1753–1792*. Uppsala: Uppsala Universitet, 2009.

Petersens, Hedvig af. "Om Torbern Bergmans och C. W. Scheeles franska förbindelser." *Personhistorisk Tidskrift* (1928).

Platen, Magnus von. "Den sörjande klienten." In M. von Platen, ed., *Klient och patron: Befordringsvägar och ståndscirkulation i det gamla Sverige*, 51–63. Stockholm: Natur och Kultur, 1988.

Porter, Roy. *The Enlightenment*. New York: Macmillan, 1990.

Porter, Roy, and Mikuláš Teich. *The Enlightenment in national context*. Cambridge: Cambridge University Press, 1981.

Porter, Theodore M. "The promotion of mining and the advancement of science: The chemical revolution of mineralogy." *Annals of Science* 38 (1981): 5439–70.

Powers, John C. *Inventing chemistry: Herman Boerhaave and the reform of the chemical arts*. Chicago and London: University of Chicago Press, 2012.

Principe, Lawrence M. "Alchemy restored." *Isis* 102 (2011): 305–12.

———. *The aspiring adept: Robert Boyle and his alchemical quest*. Princeton: Princeton University Press, 1998.

———. "Transmuting history." *Isis* 98 (2007): 779–87.

Raj, Kapil. *Relocating modern science: Circulation and the construction of knowledge in South Asia and Europe, 1650–1900*. Houndmills, Basingstoke: Palgrave Macmillan, 2007.

Roberts, Lissa. "Filling the space of possibilities: Eighteenth-century chemistry's transition from art to science." *Science in Context* 6 (1993).

Rodhe, Staffan. *Matematikens utveckling i Sverige fram till 1731*. Uppsala: Uppsala Universitet, 2002.

Rosen, Carl von, ed. *Bref från Samuel Bark till Olof Hermelin 1702–1708: Senare delen 1705–1708*. Stockholm: Norstedt, 1915.

Rudwick, Martin. *The great Devonian controversy: The shaping of scientific knowledge among gentlemanly specialists*. Chicago and London: University of Chicago Press, 1985.

Russell, Colin A. "Science on the fringe of Europe: Eighteenth-century Sweden." In D. Goodman and C. Russell, eds., *The rise of scientific Europe, 1500–1800*, 305–32. London: Hodder and Stoughton, 1991.

Rydberg, Sven. *Svenska studieresor till England under frihetstiden.* Uppsala and Stockholm: Almqvist och Wiksell, 1951.

Rydén, Göran. "The Enlightenment in practice: Swedish travellers and knowledge about the metal trades." *Sjuttonhundratal* 1 (2013).

Sächsische biographie. http://saebi.isgv.de.

Savin, Kristiina. *Fortunas klädnader: Lycka, olycka och risk i det tidigmoderna Sverige.* Lund: Sekel, 2011.

———. "Gud i Naturens laboratorium: Varsel och järtecken hos Urban Hiärne." In S. Ö. Ohlson and S. Tomingas-Joandi, eds., *Den otidsenlige Urban Hiärne: Föredrag från det internationella Hiärne-symposiet i Saadjärve, 31 augusti–4 september 2005.* Tartu: Trükikoda Greif, 2008.

Schmidt, Benjamin. "Inventing exoticism: The project of Dutch geography and the marketing of the world, circa 1700." In P. H. Smith and P. Findlen, *Merchants and marvels: Commerce, science, and art in early modern Europe,* 347–69. New York and London: Routledge, 2002.

Scribner, Robert W. "The reformation, popular magic, and the 'disenchantment of the world.'" In R. W. Scribner, *Religion and culture in Germany (1400–1800),* ed. Lyndal Roper, 346–65. Leiden: Brill, 2001.

Selling, Gösta. "Myntverkets byggnader i Stockholm." In Torsten Swensson, ed., *Kungliga myntet, 1850–1950,* 31–49. Stockholm: Nordisk Rotogravyr, 1950.

Shapin, Steven. "The invisible technician." *American Scientist* 77 (1989): 554–63.

———. *The scientific revolution.* Chicago and London: University of Chicago Press, 1996.

———. *A social history of truth: Civility and science in seventeenth-century England.* Chicago and London: University of Chicago Press, 1994.

Shapin, Steven, and Simon Schaffer. *Leviathan and the air-pump: Hobbes, Boyle, and the experimental life.* Princeton: Princeton University Press, 1985.

Siegfried, Robert, and Betty J. T. Dobbs. "Composition: A neglected aspect of the chemical revolution." *Annals of Science* 24 (1968).

Simon, Jonathan. *Chemistry, pharmacy and revolution in France, 1777–1809.* Aldershot, UK: Ashgate, 2005.

Smith, Cyril Stanley. "The discovery of carbon in steel." *Technology and Culture* 5, no. 2. (1964): 149–75.

———. "The interaction of science and practice in the history of metallurgy." *Technology and Culture* 2, no. 4 (1961): 357–67.

———. "The texture of matter as viewed by artisan, philosopher, and scientist in the seventeenth and eighteenth centuries." In Cyril Stanley Smith and John G. Burke, eds., *Atoms, blacksmiths, and crystals: Practical and theoretical views of the structure of matter in the seventeenth and eighteenth centuries,* 3–34. Los Angeles: William Andrews Clark Memorial Library, 1967.

Smith, Pamela H. *The business of alchemy: Science and culture in the Holy Roman Empire.* Princeton: Princeton University Press, 1994.

Sörlin, Per. *"Wicked arts": Witchcraft and magic trials in southern Sweden, 1635–1754.* Leiden: Brill, 1999.

Sörlin, Sverker. *De lärdas republik: Om vetenskapens internationella tendenser.* Malmö: Liber-Hermod, 1994.

Stavenow, Ludvig. *Om riksrådsvalen under frihetstiden: Bidrag till svenska riksrådets historia*. Uppsala: Almqvist och Wiksell, 1890.

Stengel, Friedemann. "Aufklärung bis zum Himmel: Emanuel Swedenborg im Kontext der Theologie und Philosophie des 18 Jahrhunderts." Habilitationsschrift der Theologischen Fakultät der Ruprecht-Karls-Universität Heidelberg. Heidelberg, 2009.

Strandberg, Olof. *Urban Hiärnes ungdom och diktning*. Stockholm and Uppsala: Almqvist och Wiksell, 1942.

Svenska akademiens ordbok. Internet ed.: http://g3.spraakdata.gu.se/saob/.

Thorndike, Lynn. *The seventeenth century*. Vol. 7 in *A history of magic and experimental science*. New York: Columbia University Press, 1958.

Tillhagen, Carl-Herman. *Järnet och människorna: Verklighet och vidskepelse*. Stockholm: LTs förlag, 1981.

Toksvig, Signe. *Emanuel Swedenborg: Scientist and mystic*. New Haven: Yale University Press, 1948.

Trofast, Jan, ed. *Johan Gottlieb Gahn Brev: Utgivna med kommentarer av Jan Trofast*. Pt. 2. Lund: Jan Trofast, 1994.

Troitzsch, Ulrich. "Kunckel von Löwenstern, Johann." In *Neue Deutsche Biographie*, 1982 ed., 13:287–88.

Vallvey, Luis Fermin Capitan. "The Spanish monopoly of Platina: Stages in the development and implementation of a policy." *Platinum Metals Review* 38 (1994): 22–25.

Venturi, Franco. *Italy and the Enlightenment: Studies in a Cosmopolitan Century*. Edited with an introduction by Stuart Woolf. London: Longman, 1972.

Vérin, Hélène. *La gloire des ingénieurs: L'intelligence technique du XVIe au XVIII siècle*. Paris: Albin Michel, 1993.

Vogel, Jakob. "Von der Wissenschafts- zur Wissensgeschichte: Für eine Historiesierung der Wissensgesellschaft." *Geschichte und Gesellschaft* 30, no. 4. (2004): 639–60.

Wakefield, Andre. *The disordered police state: German cameralism as science and practice*. Chicago and London: University of Chicago Press, 2009.

———. "Police chemistry." *Science in Context* 12, no. 3 (1999): 231–67.

Weber, Max. *Kapitalismens uppkomst: Urval och förord av Hans L. Zetterberg*. Trans. Leif Björk. Göteborg: Timbro, 1986.

Weber, Wolfhard. *Innovationen im frühindustriellen deutschen Bergbau und Hüttenwesen: Friedrich Anton von Heynitz*. Studien zu Naturwissenschaft, Technik und Wirtschaft im Neunzehnten Jahrhundert Herausgegeben von Wilhelm Treue 6. Göttingen: Vanderhoeck und Ruprecht, 1976,

Werrett, Simon. *Fireworks: Pyrotechnic arts ans sciences in European history*. Chicago and London: University of Chicago Press, 2010.

———. "An odd sort of exhibition: The St. Petersburg Academy of Sciences in enlightened Russia." Dissertation. University of Cambridge, March 2000.

Widmalm, Sven. "Instituting science in Sweden." In R. Porter and M. Teich, *The scientific revolution in national context*, 240–62. Cambridge: Cambridge University Press, 1992.

———. *Mellan kartan och verkligheten: Geodesi och kartläggning 1695–1860*. Uppsala: Uppsala Universitet, 1990.

Winton, Patrik. *Frihetstidens politiska praktik: Nätverk och offentlighet 1746–1766*. Acta Universitatis Upsaliensis Studia Historica Upsaliensia 223. Uppsala: Uppsala Universitet, 2006.

Zenzén, Nils. "Förord." In Emanuel Swedenborg, *Om järnet och de i Europa vanligaste vedertagna järnframställningssätten . . . under redaktion av Hj. Sjögren*, xiii–xvii. 1734; Stockholm: Wahlström och Widstrand, 1923.

———. "Från den tid, då vi skulle transmutera järn till koppar och få lika mycket silver i Sverige som gråberg." *Med Hammare och Fackla* 7 (1936): 88–151.

———. "Johan Gottschalk Wallerius, 1709–1785, and Axel Fredrik Cronstedt, 1722–1765." In S. Lindroth, ed., *Swedish men of science, 1650–1950*, 92–97. Stockholm: Almqvist och Wiksell, 1952.

———. "Studier i och rörande Bergskollegii Mineralsamling." *Arkiv för kemi, mineralogi och geologi* 8, no. 1 (1920): 1–134.

Index

Page numbers in italics indicate figures.